SpringerBriefs in Applied Sciences and Technology

Manufacturing and Surface Engineering

Series Editor

J. Paulo Davim

For further volumes:
http://www.springer.com/series/10623

Mark J. Jackson

Micromachining with Nanostructured Cutting Tools

 Springer

Mark J. Jackson
Saint-Gobain Abrasives
High Performance Materials
Northborough, MA
USA

ISBN 978-1-4471-4596-7   ISBN 978-1-4471-4597-4   (eBook)
DOI 10.1007/978-1-4471-4597-4
Springer London Heidelberg New York Dordrecht

Library of Congress Control Number: 2012949336

Printed on acid-free paper

Springer is part of Springer Science+Business Media (www.springer.com)

Preface

There is a substantial increase in the specific energy required with a decrease in the depth of cut during micromachining. It is believed this is due to the fact that all metals contain defects such as grain boundaries, missing and impurity atoms, substitutional atoms, defects, secondary phases, etc., and when the size of the material removed decreases the probability of encountering a stress—reducing defect decreases. Since the shear stress and strain in metal cutting is unusually high, discontinuous microcracks usually form on the primary shear plane. If the material is very brittle, or the compressive stress on the shear plane is relatively low, microcracks will grow into larger cracks giving rise to discontinuous chip formation. When discontinuous microcracks form on the shear plane they will weld and reform as strain proceeds, thus joining the transport of dislocations in accounting for the total slip of the shear plane. In the presence of a contaminant, such as carbon tetrachloride vapor at a low cutting speed, the re-welding of microcracks will decrease, resulting in a decrease in the cutting force required for chip formation. The initial part of this 'Springer Brief' briefly describes the process of micromachining and the underlying theories that describe chip formation and it shows how elementary cutting theory can be applied to machining at the microscale.

The second part shows how frictional interactions between uncoated and micro tools coated with nanostructured coatings can be characterized by using the elementary micromachining theories that were initially developed for machining at the macroscale. Shaw's methods for calculating temperatures at the interaction zone and Merchant's methods for calculating mechanical interactions are well described and justified for machining steel in both the dry and wet states. The further development and use of micro tools coated with thin-film nanostructured diamonds are shown in the third part of this 'Springer Brief'. The brief is written specifically for engineers and scientists working in this new field of micro and nanotechnology, and it explains how to characterize, apply, and adapt traditional approaches of understanding the mechanics of practical machining to the machining of microproducts using nanostructured tools.

Mark J. Jackson
Saint-Gobain Abrasives
High Performance Materials
Northborough, MA
USA

Contents

Chapter 1
Micromachining

It has been known for a long time that a size effect exists in metal cutting, where the specific energy increases with decrease in deformation size. Backer et al. [1] performed a series of experiments in which the shear energy per unit volume deformed (u_S) was determined as a function of specimen size for a ductile metal (SAE 1112 steel). The deformation processes involved were as follows, listed from top to bottom with increasing size of specimen deformed:

- Surface grinding;
- Micromilling;
- Turning; and
- Tensile test.

The surface grinding experiments were performed under relatively mild conditions involving plunge type experiments in which an 8-inch (20.3 cm) diameter wheel was directed radially downward against a square specimen of length and width 0.5 in (1.27 cm). The width of the wheel was sufficient to grind the entire surface of the work at different downfeed rates (t). The vertical and horizontal forces were measured by a dynamometer supporting the workpiece. This enabled the specific energy (u_S) and the shear stress on the shear plane (τ) to be obtained for different values of undeformed chip thickness (t). The points corresponding to a constant specific energy below a value of down feed of about 28 μ inch (0.7 μm) are on a horizontal line due to a constant theoretical strength of the material being reached when the value of, t, goes below approximately 28 μ inch (0.7 μm). The reasoning in support of this conclusion is presented by Backer et al. [1].

In the micromilling experiments, a carefully balanced 6-inch (152 cm) carbide tipped milling cutter was used with all but one of the teeth relieved so that it operated as a fly milling cutter. Horizontal and vertical forces were measured for a number of depths of cut (t) when machining the same sized surface as in grinding. The shear stress on the shear plane (τ) was estimated by a rather detailed method

M. J. Jackson, *Micromachining with Nanostructured Cutting Tools*,
SpringerBriefs in Manufacturing and Surface Engineering,
DOI: 10.1007/978-1-4471-4597-4_1, © The Author(s) 2013

presented in Backer et al. [1]. Turning experiments were performed on a 2.25-inch (5.72 cm) diameter SAE 1112 steel bar pre-machined in the form of a thin-walled tube having a wall thickness of 0.2 inch (5 mm). A zero degree rake angle carbide tool was operated in a steady-state two-dimensional orthogonal cutting mode as it machined the end of the tube. Values of shear stress on the shear plane (τ) versus undeformed chip thickness were determined for experiments at a constant cutting speed and different values of axial infeed rate and for variable cutting speeds and a constant axial infeed rate.

A true stress-strain tensile test was performed on a 0.505-inch (1.28 cm) diameter by 2-inch (5.08 cm) gage length specimen of SAE 1112 steel. The mean shear stress at fracture was 22,000 psi (151.7 MPa). Shaw [2] discusses the origin of the size effect in metal cutting, which is believed to be primarily due to short-range inhomogeneities present in all engineering metals.

When the back of a metal cutting chip is examined at very high magnification by means of an electron microscope individual slip lines. In deformation studies, Heidenreich and Shockley [3] found that slip does not occur on all atomic planes but only on certain discrete planes. In experiments on deformed aluminium single crystals the minimum spacing of adjacent slip planes was found to be approximately 50 atomic spaces while the mean slip distance along the active slip planes was found to be about 500 atomic spaces. These experiments further support the observation that metals are not homogeneous and suggest that the planes along which slip occurs are associated with inhomogeneities in the metal. Strain is not uniformly distributed in many cases. For example, the size effect in a tensile test is usually observed only for specimens <0.1 inch (2.5 mm) in diameter. On the other hand, a size effect in a torsion test occurs for considerably larger samples due to the greater stress gradient present in a torsion test than in a tensile test. This effect and several other related ones are discussed in detail by Shaw [2].

1.1 Shear Angle Prediction

There have been many notable attempts to derive an equation for the shear angle (ϕ) for steady-state orthogonal cutting. Ernst and Merchant [4] presented the first quantitative analysis. Forces acting on a chip at the tool point where: $R =$ the resultant force on the tool face, $R' =$ the resultant force in the shear plane, N_C and F_C are the components of R normal to and parallel to the tool face, N_S and F_S are the components of R' normal to and parallel to the cutting direction, F_Q and F_P are the components of R normal to and parallel to the cutting direction, and $\beta = \tan^{-1}$ F_C/N_C (is called the friction angle).

Assuming the shear stress on the shear plane (τ) to be uniformly distributed it is evident that:

$$\tau = \frac{F_S}{A_S} = \frac{R' \cos(\phi + \beta - \alpha) \sin \phi}{A} \qquad (1.1)$$

Where A_S and A are the areas of the shear plane and that corresponding to the width of cut (b), times the depth of cut (t). Ernst and Merchant [4] reasoned that τ should be an angle such that τ would be a maximum and a relationship for ϕ was obtained by differentiating Eq. 1.1 with respect to ϕ and equating the resulting expression to zero produces,

$$\phi = 45 - \frac{\beta}{2} + \frac{\alpha}{2} \tag{1.2}$$

However, it is to be noted that in differentiating, both R' and β were considered independent of ϕ.

Merchant [5] presented a different derivation that also led to Eq. 1.2. This time an expression for the total power consumed in the cutting process was first written as,

$$P = F_P V = (\tau A V) \frac{\cos(\beta - \alpha)}{\sin \phi \cos(\phi + \beta - \alpha)} \tag{1.3}$$

It was then reasoned that ϕ would be such that the total power would be a minimum. An expression identical to Eq. 1.2 was obtained when P was differentiated with respect to ϕ, this time considering τ and β to be independent of ϕ. Piispanen [6] had done this previously in a graphical way. However, he immediately carried his line of reasoning one step further and assumed that the shear stress τ would be influenced directly by normal stress on the shear plane as follows,

$$\tau = \tau_0 + K_\sigma \tag{1.4}$$

Where K is a material constant. Piispanen then incorporated this into his graphical solution for the shear angle. Upon finding Eq. 1.2 to be in poor agreement with experimental data Merchant also independently (without knowledge of Piispanen's work at the time) assumed that the relationship given in Eq. 1.4, and proceeded to work this into his second analysis as follows.

$$\sigma = \tau \tan(\phi + \beta - \alpha) \tag{1.5}$$

Or, from Eq. 1.4

$$\tau_0 = \tau + K\tau \tan(\phi + \beta - \alpha) \tag{1.6}$$

Hence,

$$\tau = \frac{\tau_0}{1 - K \tan(\phi + \beta - \alpha)} \tag{1.7}$$

When this is substituted into Eq. 1.3 we have,

$$P = \frac{\tau_0 A V \cos(\beta - \alpha)}{[1 - K \tan(\phi + \beta - \alpha)] \sin \phi \cos(\phi + \beta - \alpha)} \tag{1.8}$$

Table 1.1 Values of C in Eq. 1.9 for a variety of work and tool materials in finish turning without a cutting fluid.

Work material	Tool material	C (Degrees)
SAE 1035 Steel	HSS*	70
SAE 1035 Steel	Carbide	73
SAE 1035 Steel	Diamond	86
AISI 1022 (leaded)	HSS*	77
AISI 1022 (leaded)	Carbide	75
AISI 1113 (sulphurized)	HSS*	76
AISI 1113 (sulphurized)	Carbide	75
AISI 1019 (plain)	HSS*	75
AISI 1019 (plain)	Carbide	79
Aluminum	HSS*	83
Aluminum	Carbide	84
Aluminum	Diamond	90
Copper	HSS*	49
Copper	Carbide	47
Copper	Diamond	64
Brass	Diamond	74

Re-produced from 'micro and nanomanufacturing, compiled and Edited by M. J. Jackson, Chap. 4, 'Meso-micromachining' by M. C. Shaw and M. J. Jackson, pp. 143–190, re-printed with kind permission from Springer Science + Business Media B.V

Now, when P is differentiated with respect to ϕ and equated to zero (with τ_0 and p considered independent of ϕ we obtain,

$$\phi = \frac{\cot^{-1}(K)}{2} - \frac{\beta}{2} + \frac{\alpha}{2} = \frac{C - \beta + \alpha}{2} \tag{1.9}$$

Merchant called the quantity, $\cot^{-1} K$, the machining "constant" C. The quantity C is seen to be the angle the assumed line relating τ and ϕ makes with the τ axis [5–7]. Merchant [8] has determined the values of C given in Table 1.1 for materials of different chemistry and structure being turned under finishing conditions with different tool materials. From this table it is evident that C is not a constant. Merchant's empirical machining "constant" C that gives rise to Eq. 1.9 with values of ϕ is in reasonably good agreement with experimentally measured values.

While it is well established that the rupture stress of both brittle and ductile materials is increased significantly by the presence of compressive stress (known as the Mohr Effect), it is generally believed that a similar relationship for flow stress does not hold. However, an explanation for this paradox with considerable supporting experimental data is presented below. The fact that this discussion is limited to steady-state chip formation rules out the possibility of periodic gross cracks being involved. However, the role of micro-cracks is a possibility consistent with steady-state chip formation and the influence of compressive stress on the flow stress in shear. A discussion of the role micro-cracks can play in steady-state chip formation is presented in the next section. Hydrostatic stress plays no role in the plastic flow of metals if they have no porosity. Yielding then occurs when the von Mises criterion reaches a critical value. Merchant [7] has indicated that Barrett [9] found that for single crystal metals τ_S is independent of ϕ when plastics such as celluloid are cut. In general, if a small amount of compressibility is involved yielding will occur when the von Mises criterion reaches a certain value.

However, based on the results of Table 1.1 the role of compressive stress on shear stress on the shear plane in steady-state metal cutting is substantial. The fact there is no outward sign of voids or porosity in steady-state chip formation of a ductile metal during cutting and yet there is a substantial influence of normal stress on shear stress on the shear plane represents an interesting paradox. It is interesting to note that Piispanen [6] had assumed that shear stress on the shear plane would increase with normal stress and had incorporated this into his graphical treatment.

There is much research in the area of the analysis of micromachining from an experimental and computational viewpoint. From an experimental viewpoint, Robinson [10] has conducted very thorough analysis in his dissertation that looks at applying the earlier metal cutting theories developed by Ernst and Merchant [4], Merchant [7, 8] and Shaw [2] to micromachining a variety of different materials such as pure metals and steels of varying carbon content. Novakov [11] analysed the machining of titanium and its alloys from a computational perspective and developed an understanding of the mechanics of chip formation at the microscale. Other researchers have published papers in this exciting and the main publications are shown in the reference section of this book. The next chapter discusses the nature of contact between the micro cutting tool and the workpiece in a atypical microcutting application and demonstrated the applicability of using Ernst and Merchant's, Merchant's and Shaw's models to describe the metal cutting process at the microscale.

Acknowledgments The author acknowledges permission to reproduce parts of this chapter from the following publication: 'Micro and Nanomanufacturing, Compiled and Edited by M. J. Jackson, Chap. 4, 'Meso-micromachining' by M. C. Shaw and M. J. Jackson, pp. 143–190, re-printed with kind permission from Springer Science + Business Media B.V. (Permission Received February 22, 2012).

References

1. Backer WR, Marshall ER, Shaw MC (1952) Trans ASME 74:61
2. Shaw MC (1952) J Franklin Inst 254(2):109
3. Heidenreich RO, Shockley W (1948) Report on strength of solids. Phys, Society of London 57
4. Ernst HJ, Merchant ME (1941) Trans Am Soc Metals 29:299
5. Merchant ME (1945) J Appl Phys 16:267–275
6. Piispanen V (1937) Teknillinen Aikakaushehti (Finland) 27:315
7. Merchant ME (1945) J Appl Phys 16:318–324
8. Merchant ME (1950) Machining theory and practice. Am Soc Metals 9:5–44
9. Barrett CS (1943) Structure of metals. McGraw Hill Co, NY 295
10. Robinson GM (2007) Doctoral dissertation, Purdue University
11. Novakov T (2010) Doctoral dissertation, Purdue University

Chapter 2
Analysis of Contact Between Chip and Tool Using Nanostructured Coated Cutting Tools

2.1 Introduction

The stages of contact between a metal chip of AISI 1018 steel and a coated cutting tool creates significant opportunities for manufacturers of machined products to understand how dry machining and minimum quantity lubrication affects the economics of manufacturing. The present work not only compares various computational approaches to the solution of shear plane and tool face temperatures during dry machining with nanostructured coated milling tools, but also explains why there is a large discrepancy when calculating temperature generated during machining when using Loewen and Shaw's method for calculating shear plane and tool face temperatures. There has been a great deal of activity in understanding metal cutting mechanics. Traditional metal cutting theories are being challenged as a result of advances in computational mechanics. Astakhov and co-workers [1–4] are particularly critical of using theories developed in the 1940s that describe the mechanics of metal cutting. In response to Astakhov's [1–4] assertions that the previous theories do not readily apply to current machining practices, the authors have conducted a series of computational analyses of machining AISI 1018 steel in order to understand if the Loewen and Shaw's method can be applied to calculating shear plane and tool face temperatures by comparing with finite element models that were constructed using a commercial software package known as Third Wave Systems^{TM}. The following analysis allows one to test if Astakhov's statements apply to the primary stages of chip formation and the subsequent stages of chip formation where a secondary shear zone is established that is caused by the formation of a built-up edge [5–14].

M. J. Jackson, *Micromachining with Nanostructured Cutting Tools*,
SpringerBriefs in Manufacturing and Surface Engineering,
DOI: 10.1007/978-1-4471-4597-4_2, © The Author(s) 2013

2.2 Computational Analysis of Machining Conditions

2.2.1 Loewen and Shaw's Method to Calculating Cutting Temperatures

There are several analytical approaches used to determine the cutting temperature. These methods are suitable for the analysis of soft materials, in particular low carbon steels containing a high percentage of ferrite. Of these methods, Stephenson [13] found that the most accurate model was Loewen and Shaw's because it accounted for the change in thermal properties of the tool and workpiece with temperature. Therefore, Loewen and Shaw's approach is used as demonstrated by Shaw [12] in order to explain the use of such formulae for the current experimental work. However, since there is no dynamometer currently available to measure cutting forces, the horizontal force can be closely approximated by the formula proposed by Isakov [8]. Once this quantity has been determined, Loewen and Shaw's approach can be used. Isakov's formula to find the tangential force or the force, in the horizontal orientation, F_{HO}, of a milling cutter is given by,

$$F_{HO} = \sigma_{UTS} A n_c C_m C_w \qquad (2.1)$$

where, σ_{UTS}, is the ultimate tensile strength (UTS) of the workpiece (for AISI 1018 steel, $\sigma = 4.18 \times 10^{10}$ MPa), A, is the uncut chip cross sectional area, n_c, is the number of teeth engaged in the workpiece, C_m, is a machinability adjustment factor, and C_w, is a tool wear adjustment factor. The feed per tooth, f_t, must be calculated, to do this the following quantities must be known, the feed, $f = 0.381$ m/min $= 0.00635$ m/s, the spindle speed, N $= 2500$ rpm, or in radians the angular velocity, $\omega = 261.8$ rad/s, the number of cutting teeth, n $= 4$. The feed per tooth can now be calculated from the following equation,

$$f_t = \frac{f}{Nn} \qquad (2.2)$$

$$f_t = \frac{0.381 (m/min)}{2500 (rpm) \times 4} \qquad (2.3)$$

$$f_t = 3.81 \times 10^{-5} m$$

The rake angle is measured at $\alpha = 9°$, and the feed per tooth correction factor, f_{tc}, is given by,

$$f_{tc} = \frac{f_t}{Cos\ \alpha} \qquad (2.4)$$

$$f_{tc} = \frac{3.81 \times 10^{-5} (m/min)}{Cos\ 9°} \qquad (2.5)$$

$$f_{tc} = 3.86 \times 10^{-5}\ m/min$$

The method being followed is for orthogonal cutting, the case being considered is milling. Therefore, the orthogonal width of cut is replaced by the milling axial depth of cut, b, which is 1.27×10^{-3} m and the uncut chip cross sectional, A, is given by,

$$A = f_{tc}b \tag{2.6}$$

$$A = 3.86 \times 10^{-5}(m) \times 1.27 \times 10^{-3}(m) \tag{2.7}$$

$$A = 4.9 \times 10^{-8}\,m^2$$

The tool diameter, $d = 1.27 \times 10^{-2}$ m, and the width of cut, W, is 0.635×10^{-2} m. The number of teeth engaged in the cut, n_c, can now be calculated by,

$$n_c = \frac{n(90 + \sin^{-1}((2W - d)/d))}{360} \tag{2.8}$$

$$n_c = \frac{4(90 + \sin^{-1}((2 \times 0.00635(m) - 0.0127(m))/0.0127(m)))}{360} \tag{2.9}$$

$$n_c = 1$$

The machinability adjustment factor taken from Isakov's [8] textbook and has a value of unity, and the tool wear adjustment factor is taken as 1.1, also from Isakov [8]. Therefore, F_{HO} can be calculated.

$$F_{HO} = \sigma A n_c C_m C_w \tag{2.10}$$

$$F_{HO} = 4.2 \times 10^8 \times 4.9 \times 10^{-8} \times 1 \times 1 \times 1.1 \tag{2.11}$$

$$F_{HO} = 22.63\,N$$

The chip thickness, t_c, was measured to be 1×10^{-4} m, thus, the chip thickness ratio, r, can be calculated from,

$$r = \frac{t_o}{t_c} \tag{2.12}$$

$$r = \frac{3.81 \times 10^{-5}(m)}{1 \times 10^{-4}(m)} \tag{2.13}$$

$$r = 0.381$$

This allows the shear plane angle, Φ, to be calculated from,

$$\Phi = \tan^1\left(\frac{rCos\alpha}{1 - rSin\alpha}\right) \tag{2.14}$$

$$\Phi = \tan^1\left(\frac{0.381Cos9°}{1 - 0.381Sin9°}\right) \tag{2.15}$$

$$\Phi = 21.81°$$

Based on the work of Bowden and Tabor [7] the coefficient of friction between tungsten carbide and a low carbon steel (in this case AISI 1018 steel) during dry rubbing conditions is, $\mu = 0.78$. This allows the calculation of the force in the vertical orientation F_{VO}, which is given by,

$$F_{VO} = \frac{\mu F_{HO} - F_{HO}\text{Tan}\,\alpha}{1 + \mu\text{Tan}\,\alpha} \tag{2.16}$$

$$F_{VO} = \frac{0.78 \times 22.63(\text{N}) - 22.63(\text{N})\text{Tan}\,9°}{1 + 0.78\text{Tan}\,9°} \tag{2.17}$$

$$F_{VO} = 12.52\,\text{N}$$

The force along the tool face, F_{AT}, is given by,

$$F_{AT} = F_{HO}\text{Sin}\,\alpha + F_{VO}\text{Cos}\,\alpha \tag{2.18}$$

$$F_{AT} = 22.63(\text{N})\text{Sin}\,9° + 12.52(\text{N})\text{Cos}\,9° \tag{2.19}$$

$$F_{AT} = 15.91\,\text{N}$$

The force normal to the tool face, F_{NT}, can be calculated by,

$$F_{NT} = F_{HO}\text{Cos}\,\alpha - F_{VO}\text{Sin}\,\alpha \tag{2.20}$$

$$F_{NT} = 22.63(\text{N})\text{Cos}\,9° - 12.52(\text{N})\text{Sin}\,9° \tag{2.21}$$

$$F_{NT} = 20.4\,\text{N}$$

The force along the shear plane, F_{AS}, is given by,

$$F_{AS} = F_{HO}\text{Cos}\,\Phi - F_{VO}\text{Sin}\,\Phi \tag{2.22}$$

$$F_{AS} = 22.63(\text{N})\text{Cos}\,21.81° - 12.52(\text{N})\text{Sin}\,21.81° \tag{2.23}$$

$$F_{AS} = 16.36\,\text{N}$$

The force normal to the shear plane, F_{NS}, is given by,

$$F_{NS} = F_{VO}\text{Cos}\,\Phi + F_{HO}\text{Sin}\,\Phi \tag{2.24}$$

$$F_{NS} = 12.52(\text{N})\text{Cos}\,21.81° + 22.63(\text{N})\text{Sin}\,21.81° \tag{2.25}$$

$$F_{NS} = 20.03\text{N}$$

Since the milling process is being approximated by an orthogonal cutting operation, the milling axial depth of cut is equal to the orthogonal chip width, b. The maximum uncut chip thickness is equal to the feed per tooth t_o. The area of the shear plane A_s is given by,

$$A_s = \frac{b t_O}{Sin\, \Phi} \tag{2.26}$$

$$A_s = \frac{(1.27 \times 10^{-3}(m))(3.81 \times 10^{-5}(m))}{Sin\, 21.81^\circ} \tag{2.27}$$

$$A_s = 1.3 \times 10^{-7}\, m^2$$

The shear stress, τ, is given by,

$$\tau = \frac{F_{AS}}{A_S} \tag{2.28}$$

$$\tau = 16.36(N)/1.3 \times 10^{-7}(m^2) \tag{2.29}$$

$$\tau = 125.62 \times 10^6\, N/m^2$$

Similarly the normal stress, σ, is given by,

$$\sigma = \frac{F_{NS}}{A_S} \tag{2.30}$$

$$\sigma = 20.03(N)/1.3 \times 10^{-7}(m^2) \tag{2.31}$$

$$\sigma = 153.82 \times 10^6\, N/m^2$$

The shear strain, γ, is given by,

$$\gamma = \frac{Cos\, \alpha}{Sin\, \Phi\, Cos(\Phi - \alpha)} \tag{2.32}$$

$$\gamma = \frac{Cos\, 9^\circ}{Sin\, 21.81^\circ Cos(21.81^\circ - 9^\circ)} \tag{2.33}$$

$$\gamma = 2.73$$

The cutting velocity, V, at the tool tip is given by,

$$V = r_t \omega$$

where, r_t, is the tool radius and the units of, ω, are radians/s.

$$V = 0.00635(m)\left(\frac{2\pi}{60(s)} \times 2500(rpm)\right) \tag{2.34}$$

$$V = 1.66\, m/s$$

The chip velocity, V_C, is given by,

$$V_C = \frac{V\, Sin\, \Phi}{Cos(\Phi - \alpha)} \tag{2.35}$$

$$V_C = \frac{1.66(\text{m/s})\text{Sin}\,21.81°}{\text{Cos}(21.81° - 9°)} \qquad (2.36)$$

$$V_C = 0.632\,\text{m/s}$$

Similarly the shear velocity, V_S, is given by,

$$V_s = \frac{V\,\text{Cos}\,\alpha}{\text{Cos}(\Phi - \alpha)} \qquad (2.37)$$

$$V_s = \frac{1.66(\text{m/s})\text{Cos}\,9°}{\text{Cos}(21.81° - 9°)} \qquad (2.38)$$

$$V_s = 1.68\,\text{m/s}$$

In order to determine the strain rate, $\dot{\gamma}$, the shear plane spacing, Δy, must be determined from chip images, in this case $\Delta y = 10\ \mu\text{m}$.

$$\dot{\gamma} = \frac{V\,\text{Cos}\,\alpha}{\Delta y\,\text{Cos}(\Phi - \alpha)} \qquad (2.39)$$

$$\dot{\gamma} = \frac{1.66(\text{m/s})\text{Cos}\,9°}{(1 \times 10^{-5}(\text{m}))\text{Cos}(21.81° - 9°)} \qquad (2.40)$$

$$\dot{\gamma} = 168.39 \times 10^3\,\text{s}^{-1}$$

The theoretical scallop height, h, which reflects the surface roughness of the machined surface in an end milling operation is given by,

$$h = \frac{f_t^2}{4d} \qquad (2.41)$$

$$h = \frac{\left(3.81 \times 10^{-5}(\text{m})\right)^2}{4(0.0127(\text{m}))} \qquad (2.42)$$

$$h = 2.86 \times 10^{-8}\,\text{m}$$

The energy per unit time, U, is given by,

$$U = F_{HO}V$$

$$U = 22.63(\text{N}) \times 1.66(\text{m/s})$$

$$U = 37.63\,\text{Nm}\,\text{s}^{-1},\ \text{or}\ 37.63\,\text{J}\,\text{s}^{-1}$$

The energy per unit volume, u, is given by,

$$u = \frac{F_{HO}}{bt_0} \qquad (2.43)$$

$$u = \frac{22.63(\text{N})}{(1.27 \times 10^{-3}(\text{m}))(3.81 \times 10^{-5}(\text{m}))} \tag{2.44}$$

$$u = 467.76 \times 10^6 \, \text{Nm/m}^3 \text{ or } 467.76 \times 10^6 \, \text{J/m}^3$$

The shear energy per unit volume, u_s, is given by,

$$u_s = \tau\gamma \tag{2.45}$$

$$u_s = 125622347(\text{N/m}^2) \times 2.73(\text{m/m}) \tag{2.46}$$

$$u_s = 342.49 \times 10^6 \, \text{Nm/m}^3 \text{ or } 342.49 \times 10^6 \, \text{J/m}^3$$

The friction energy per unit volume, u_f, is given by,

$$u_f = \frac{F_{AT}r}{bt_0} \tag{2.47}$$

$$u_f = \frac{(15.91(\text{N}))(0.381)}{(1.27 \times 10^{-3}(\text{m}))(3.81 \times 10^{-6}(\text{m}))} \tag{2.48}$$

$$u_f = 125.27 \times 10^6 \, \text{Nm/m}^3 \text{ or } 125.27 \times 10^6 \, \text{J/m}^3$$

To determine the shear plane temperature, the method of Loewen and Shaw taken from Shaw [12] is used. The process involves several iterations, the final iteration is shown. The initial step is to estimate the shear plane temperature, $\theta_s = 70\,°\text{C}$, and the ambient temperature is $\theta_o = 25\,°\text{C}$, then calculate the mean of these two temperatures, θ_{av}.

$$\theta_{AV} = \frac{\theta_s + \theta_o}{2} \tag{2.49}$$

$$\theta_{AV} = \frac{70\,°\text{C} + 25\,°\text{C}}{2} \tag{2.50}$$

$$\theta_{AV} = 47.5\,°\text{C}$$

The thermal properties of the workpiece must be determined; in this case a low carbon steel. Shaw [12] displays these properties. At 47.5 °C, the thermal diffusivity, $K_1 = 1.58 \times 10^{-5} \, \text{m}^2/\text{s}$, and the volumetric specific heat is, $\rho_1 C_1 = 3.72 \times 10^6 \, \text{J/m}^3 \, °\text{C}$. Calculate the quantity R_1 from,

$$R_1 = \frac{1}{1 + 1.328\left(\frac{K_1\gamma}{Vt_0}\right)^{1/2}} \tag{2.51}$$

$$R_1 = \frac{1}{1 + 1.328\left(\frac{1.58 \times 10^{-5}(\text{m}^2/\text{s}) \times 2.73}{1.66(\text{m/s}) \times 3.81 \times 10^{-5}(\text{m})}\right)^{1/2}} \tag{2.52}$$

$$R_1 = 0.48$$

Calculate the quantity $\theta s - \theta o$ from,

$$\theta s - \theta o = \frac{R_1 u_s}{J \rho_1 C_1} \tag{2.53}$$

Where J is the mechanical equivalent of heat, the value used by Shaw [12] of 9340 lbin/BTUs^2 for low carbon steel, or 0.99 Nm/J s^2, will be used here,

$$\theta s - \theta o = \frac{0.48\left(342.49 \times 10^6 (\text{J/m}^3)\right)}{\left(0.99(\text{Nm/Js}^2)\right)\left(371.88 \times 10^4 (\text{J/m}^3\,^\circ\text{C})\right)} \tag{2.54}$$

$$\theta s - \theta o = 43.99\,^\circ\text{C}$$

Therefore,

$$\theta s = 68.99\,^\circ\text{C}$$

Shaw states the process is repeated until the initial estimate produces a shear plane temperature within 3.88 °C of the initial estimate. Therefore, in this case the accepted shear plane temperature is 68.99 °C.

To calculate the tool face temperature estimate the tool face temperature $\theta_T = 89.25$ °C. Then determine the thermal properties of the workpiece at this temperature, in this case a low carbon steel, they are taken from Shaw [12]. At 89.25 °C, $K_1 = 1.49 \times 10^{-5}$ m^2/s, $\rho_1 C_1 = 3.91 \times 10^6$ J/m^3 °C. The chip contact length, a, is approximated as half the uncut chip length l, which is given by,

$$l = (dt_o)^{0.5} + \frac{f}{2nN} \tag{2.55}$$

$$l = \left(0.0127(\text{m}) \times 3.81 \times 10^{-3}(\text{m})\right)^{0.5} + \frac{0.381(\text{m})}{2 \times 4 \times 2500(\text{rpm})} \tag{2.56}$$

$$l = 7.15 \times 10^{-4}\,\text{m}$$

Thus,

$$a = l/2 \tag{2.57}$$

$$a = 7.15 \times 10^{-4}(\text{m})/2 \tag{2.58}$$

$$a = 3.57 \times 10^{-4}\,\text{m}$$

$$\frac{m}{l} = \frac{b}{2a} \tag{2.59}$$

$$\frac{m}{l} = \frac{1.27 \times 10^{-3}(\text{m})}{2(3.57 \times 10^{-4}(\text{m}))} \tag{2.60}$$

$$\frac{m}{l} = 8.89$$

Using Shaw's textbook [12] to find $\overline{A} = 2.1$, determine the thermal conductivity of the workpiece, k_S, at the previously calculated shear plane temperature of 68.99 °C, in this case, $k_T = 6.12 \times 10^{-3}$ J/m^2 s °C. This allows C′ to be calculated from,

$$C' = \frac{u_f V t_0 \overline{A}}{J k_T} \tag{2.61}$$

$$C' = \frac{(125.27 \times 10^6 (\text{J/m}^3))(1.66(\text{m/s}))(3.81 \times 10^{-5}(\text{m})) \times 2.1}{(0.99(\text{Nm/Js}^2))(6.12 \times 10^{-3}(\text{J/m}^2\text{s°C}))} \tag{2.62}$$

$$C\prime = 27.24 \times 10^5$$

Calculate B′ from,

$$B' = \left(\frac{0.754 u_f}{J \rho_S C_S}\right)\left(\frac{V t_o^2}{a r K_S}\right)^{1/2} \tag{2.63}$$

where the subscript S denotes the thermal properties determined at the previously calculated shear plane temperature, $\rho_S c_S = 3.91 \times 10^6$ J/m^3 °C and $K_S = 1.49 \times 10^{-5}$ m^2/s.

$$B' = \left(\frac{0.754(125.2 \times 10^6 (\text{J/m}^3))}{(0.99(\text{Nm/Js}^2))(3.91 \times 10^6 (\text{J/m}^3\text{°C}))}\right)$$
$$\times \left(\frac{(1.66(\text{m/s}))(3.81 \times 10^{-5}(\text{m}))^2}{(3.57 \times 10^{-4}(\text{m})) \times 0.381 \times (1.49 \times 10^{-5}(\text{m}^2/\text{s}))}\right)^{1/2} \tag{2.64}$$

$$B' = 18.87$$

Calculate the quantity R_2 from,

$$R_2 = \frac{C' - \theta_S + \theta_o}{C' + B'} \tag{2.65}$$

$$R_2 = \frac{(2.72 \times 10^6) - (68.99(\text{°C})) + (25(\text{°C}))}{(2.72 \times 10^6(\text{°C})) + 18.87} \tag{2.66}$$

$$R_2 = 0.99 \text{°C}$$

Calculate the temperature rise in the chip surface due to friction, $\Delta\theta_F$, from,

$$\Delta\theta_F = R_2 B' \tag{2.67}$$

$$\Delta\theta_F = 0.99(\text{°C}) \times 18.87 \tag{2.68}$$

$$\Delta\theta_F = 18.67 \text{°C}$$

Table 2.1 Table of computational parameters used for machining simulations

Cutting speed—V (m/s)	1.662426113
Undeformed chip thickness—t_0 (m)	0.0000381
Width of cut—b (m)	0.00635
Cutting force—F_p (N)	22.63344932
Feed force—F_q (N)	12.52230092
Chip thick ratio—r	0.381
Shear strain—γ	2.726369114
Shear energy/unit volume—u_s (mN/m^3)	342492886.91
Friction energy/unit volume—u_f (mN/m^3)	125266001.2
Mechanical equivalent of heat—J (Nm/Js^2)	0.998996696
Tool diameter—d (m)	0.0127
Cutting speed—V (m/min)	99.74556675
Number of cutting teeth—n	4
Cutting speed—w (rpm)	2500

Re-produced from 'Machining with Nanomaterials, Edited by M. J. Jackson and J. S. Morrell, with kind permission from Springer Science+Business Media B.V. (Permission Received February 22, 2012)

Finally the tool face temperature, θ_T, can be calculated from,

$$\theta_T = \theta_S + \Delta\theta_F \tag{2.69}$$

$$\theta_T = 68.99(^\circ C) + 18.67(^\circ C) \tag{2.70}$$

$$\theta_T = 87.66\,^\circ C.$$

Shaw [12] states that the procedure for calculating the tool face temperature should be repeated until the initial estimate and the final calculated temperature are in agreement. In this case the difference of 1.59 °C cannot be doubted since the material properties are determined from interpretation of line diagrams. Therefore, 89.25 °C is the accepted tool face temperature. Tables 2.1, 2.2 and 2.3 show the experimental parameters and equations that were used to calculate the shear plane and tool face temperatures. The tables include sample calculations to demonstrate the use of the equations and parameters shown above. Table 2.4 shows the shear

Table 2.2 Calculation of shear plane temperature

To calculate shear plane temperature—°C		
Step 1		
Estimate initial shear plane temperature θ_s		70.00 °C
Ambient temperature θ_0		25.00 °C
Average temperature		47.50 °C
Find material properties from fig 12.25 at		47.50 °C
	K_1—thermal diffusivity	1.5812E−05 m²/s
	$C_1\rho_1$—volume specific heat	3718849.84 J/m³ °C
Calculate R_1	$R_1 = 1/(1 + 1.328(K_1\gamma/Vt_0)^{1/2})$	
	$R_1 =$	0.477191655
Step 2	$\theta_s - \theta_0 = R_1 u_s/JC_1\rho_1$	
	$\theta_s - \theta_0 =$	43.99179692 °C
Since $\theta_0 =$		25.00 °C
$\theta_s =$		68.99 °C
Step 3		
If initial guess is within 3.88 °C use that temperature, if not repeat		
		−1.01 °C
	Feed rate—f	
	(m/min)	0.381
	Chip contact length—l	
	(m)	7.15E−04
	Chip contact length—a (m)	0.000357329
	(m)	

Re-produced from 'Machining with Nanomaterials, Edited by M. J. Jackson and J. S. Morrell, with kind permission from Springer Science+Business Media B.V. (Permission Received February 22, 2012)

plane approximation and calculation temperatures. It also includes the tool face approximations and calculations of the temperatures. The last column shows the percentage error between estimated and calculated temperatures. The coefficient of friction of 0.78 has been taken from Robinson [11] in order to compare with the calculations shown above. It is intended to be factual information based on the experimental work performed by Robinson [11]. The shear plane temperature and the tool face temperature have been calculated using different values of coefficient of friction in order to provide a parametric analysis and to compare with the values for the different coatings. This ranges from 0.1 to 1.0 with 0.1 increments. This has then been compared to the approximate amounts of shear plane and tool face temperatures. The percentage error is then calculated for a comparison between the approximation and the calculation. The percentage of error is less than 3.5 % for all calculations. From Figs. 2.1 and 2.2, it can be seen that increasing the coefficient of friction lowers the shear plane temperature, but conversely the tool face temperature tends to increase.

Table 2.3 Calculation of tool face temperature

To calculate tool face temperature—°C		
Step 4	Estimate initial shear plane temperature θ_T	89.25 °C
	Ambient temperature θ'_0	25 °C
Material properties from fig 12.25 [12] at estimated tool temperature of		89.25 °C
	K_2—thermal conductivity	1.49E−05 m^2/s
	$C_2\rho_2$—volume specific heat	3912751.678 J/m^3 °C
Calculate the ratio		m/l = b/2a
	m/l =	8.89
Use figure 12.17 [12] to find A		2.1
Find k_3 from fig 12.25 at calculated shear plane temperature of		68.99
	k_3	6.12E−03 J/ms °C
Step 5	Calculate the quantity	$C' = u_f V t_0 A / J k_3$
	$C' =$	2724338.49 °C
Step 6	Calculate the quantity B'	
	$B' = (0.754 u_f / J \rho_3 C_3)(V t_0^2 / a r K_2)^{1/2}$	18.66628618 °C
Step 7	Determine R_2	$R_2 = (C' - \theta_s + \theta'_0)/(C' + B')$
	$R_2 =$	0.999977001
Step 8	Temp rise in chip due to friction, $\Delta\theta_F$	$\Delta\theta_F = R_2 B'$
	$\Delta\theta_F =$	18.66585687 °C
Step 9	Calculate θ_T from	$\theta_T = \theta_s + \Delta\theta_F$
	$\theta_T =$	87.66 °C
Repeat analysis until calculated value and estimate match		
	Diff =	1.59

2.3 Finite Element Studies of Machining Conditions

A metal cutting finite element software was chosen to simulate the metal cutting operation. The choice of the software is important because the output results may vary from software to software and with the input parameters also. Therefore, AdvantEdge^{TM} (provided by Third Wave Systems) was used in this study. The software uses adaptive meshing to improve the quality and precision of the output results and it also includes a wide range of workpiece material library. The simulations are replicas of what was done experimentally and the cutting parameters used were the same in both processes. Third Wave AdvantEdge^{TM} allows improving and optimizing of machining processes. To model the thermal-visco plastic behavior of the materials, the software employs a constitutive equation, the Johnson–Cook law, which can be represented by the following formula:

$$\sigma_{eq} = (A + B\varepsilon^n)\left(1 + C\ln\left(\frac{\dot{\varepsilon}}{\dot{\varepsilon}_0}\right)\right)\left(1 - \left(\frac{T - T_{room}}{T_m - T_{room}}\right)^m\right) \qquad (2.71)$$

Table 2.4 Approximate and calculated tool face temperatures and shear plane temperatures

CoF	Shear plane approximate (°C)	Shear plane calculated (°C)	Error (°C)	Tool face approximate (°C)	Tool face calculated (°C)	Error (°C)
0.1	80	82.03	−2.03	85	87.4	−2.4
0.2	80	79.85	0.15	87	85.06	1.94
0.3	79	78	1	85	85.51	−0.51
0.4	76	75.45	0.55	84	85.32	−1.32
0.5	75	74.14	0.86	86	86.6	−0.6
0.6	74	72.71	1.29	87	87.44	−0.44
0.7	72	70.41	1.59	88	87.2	0.8
0.78	70	68.99	1.01	89.25	87.66	1.59
0.9	68	67.1	0.9	91	88.28	2.72
1	65	65.42	−0.42	92	88.58	3.42

Re-produced from 'Machining with Nanomaterials, Edited by M. J. Jackson and J. S. Morrell, with kind permission from Springer Science+Business Media B.V. (Permission Received February 22, 2012)

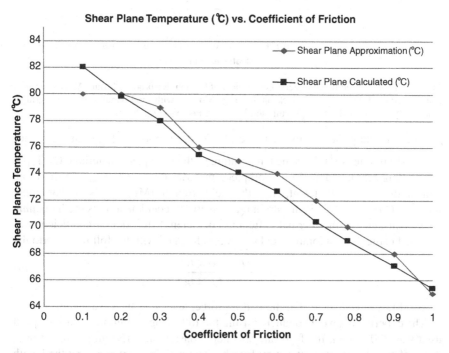

Fig. 2.1 Shear plane temperature versus coefficient of friction. Re-produced from 'Machining with Nanomaterials, Edited by M. J. Jackson and J. S. Morrell, with kind permission from Springer Science+Business Media B.V. (Permission Received February 22, 2012)

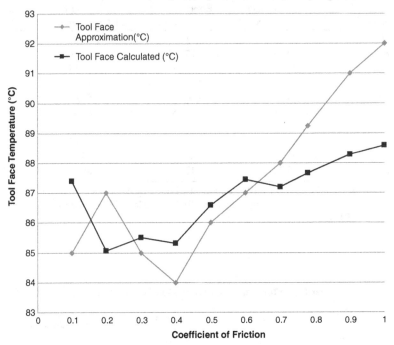

Fig. 2.2 Tool face temperature versus coefficient of friction. Re-produced from 'Machining with Nanomaterials, Edited by M. J. Jackson and J. S. Morrell, with kind permission from Springer Science+Business Media B.V. (Permission Received February 22, 2012)

where, ε, is the plastic strain, $\dot{\varepsilon}$, is the plastic strain rate (s^{-1}), $\dot{\varepsilon}_0$ is the reference plastic strain rate (s^{-1}), T, is the temperature of the workpiece material (°C), T_m, is the melting temperature of the workpiece material (°C), and, T_{room}, is the room temperature (°C). Coefficient, A, is the yield strength (MPa), B, is the hardening modulus (MPa), and, C, is the strain rate sensitivity coefficient, n, is the hardening coefficient and, m, the thermal softening coefficient. The friction coefficient was obtained using the Coulomb model and was calculated with the following formula:

$$u = \frac{Ff + Fc \times tg\gamma}{Fc - Ff \times tg\gamma}$$
(2.72)

where, Ff, the feed force, Fc, the cutting force, and, γ, is the rake angle.

The experiments were conducted using the following conditions: Cutting speed used was 300 m/min; the feed rate was kept constant at 0.15 mm per revolution; depth of cut was constant at 1 mm; length of cut was constant at 6 mm; the length of the work piece was 5 mm; the experimental specimen was 1018 steel; initial temperature was 20 °C, and the coating thickness was constant at 4 μ' thickness. The coefficient of friction was varied between 0.1 and 1 in steps of 0.1 in order to compare results generated using Loewen and Shaw's method.

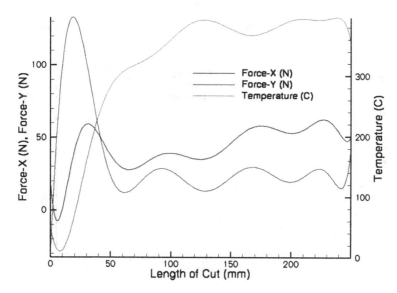

Fig. 2.3 Peak tool temperature as a function of the length of cut for a coefficient of friction of 0.1 between tool and chip. Re-produced from 'Machining with Nanomaterials, Edited by M. J. Jackson and J. S. Morrell, with kind permission from Springer Science+Business Media B.V. (Permission Received February 22, 2012)

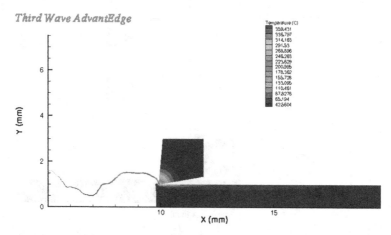

Fig. 2.4 Temperature profile between chip and tool for a coefficient of friction of 0.1 between chip and tool. Re-produced from 'Machining with Nanomaterials, Edited by M. J. Jackson and J. S. Morrell, with kind permission from Springer Science+Business Media B.V. (Permission Received February 22, 2012)

The finite element generated results (Figs. 2.3–2.8) show the machining variables peak tool temperature and machining temperatures as a function of the coefficient of friction. In order to minimize computational effort, calculations for coefficient of friction of 0.1, 0.5, and 1.0 are shown in this paper. Figures 2.9 and 2.10 shows the

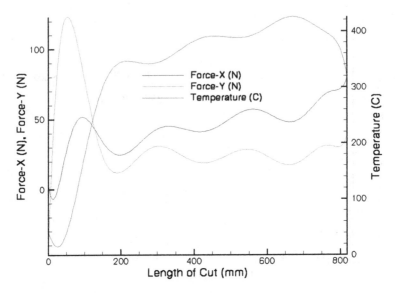

Fig. 2.5 Peak tool temperature as a function of the length of cut for a coefficient of friction of 0.5 between tool and chip. Re-produced from 'Machining with Nanomaterials, Edited by M. J. Jackson and J. S. Morrell, with kind permission from Springer Science+Business Media B.V. (Permission Received February 22, 2012)

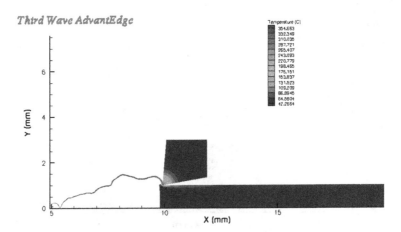

Fig. 2.6 Temperature profile between chip and tool for a coefficient of friction of 0.5 between chip and tool. Re-produced from 'Machining with Nanomaterials, Edited by M. J. Jackson and J. S. Morrell, with kind permission from Springer Science+Business Media B.V. (Permission Received February 22, 2012)

comparison between calculated, approximate, and finite element generated shear plane and tool face temperatures. It is shown that as machining develops, shear plane and tool face temperatures increase rapidly after the initial frictional interactions between workpiece and cutting tool. Here, the method of Loewen and Shaw does not

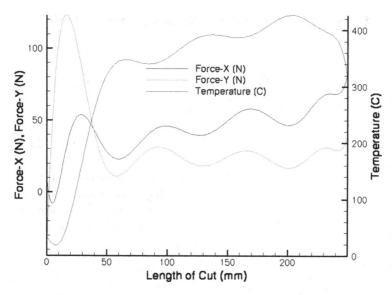

Fig. 2.7 Peak tool temperature as a function of the length of cut for a coefficient of friction of 1.0 between tool and chip. Re-produced from 'Machining with Nanomaterials, Edited by M. J. Jackson and J. S. Morrell, with kind permission from Springer Science+Business Media B.V. (Permission Received February 22, 2012)

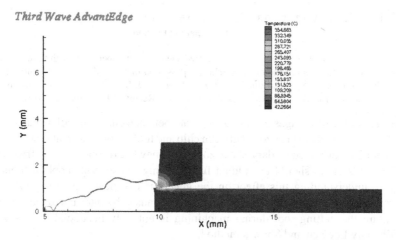

Fig. 2.8 Temperature profile between chip and tool for a coefficient of friction of 1.0 between chip and cutting tool. Re-produced from 'Machining with Nanomaterials, Edited by M. J. Jackson and J. S. Morrell, with kind permission from Springer Science+Business Media B.V. (Permission Received February 22, 2012)

compare well. However, at the beginning of the cut, i.e., after the first 25 mm is machined, it can be seen that Loewen and Shaw's method produces a very accurate estimate of shear plane and tool face temperature. This implies that their method is

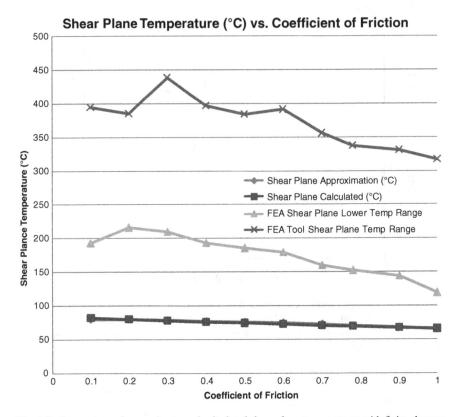

Fig. 2.9 Comparison of approximate and calculated shear plane temperatures with finite element generated temperatures at the end of the machining pass (800 mm machined). Re-produced from 'Machining with Nanomaterials, Edited by M. J. Jackson and J. S. Morrell, with kind permission from Springer Science+Business Media B.V. (Permission Received February 22, 2012)

applicable at the first stages of intimate contact between chip and tool. Clearly, it is noted that a secondary shear zone between chip and tool is not created at first contact. This implies that the secondary shear zone is somewhat responsible for frictional heating. This effect should be studied further using transparent sapphire tools in terms of understanding this effect on the generation of heat. The computational analysis leads us believe that intermittent contact caused by interrupted cuts and/or oscillating the cutting tool during machining is not only beneficial, but is well described by Loewen and Shaw's method.

2.4 Discussion

When one inspects the results of finite element modeling (Figs. 2.3–2.8), it is observed that the change in the coefficient of friction tends to lower the shear plane temperature and increase the tool face temperature as it increases in value. In order to make the removal of the chip easier, it is advisable to focus the generation of

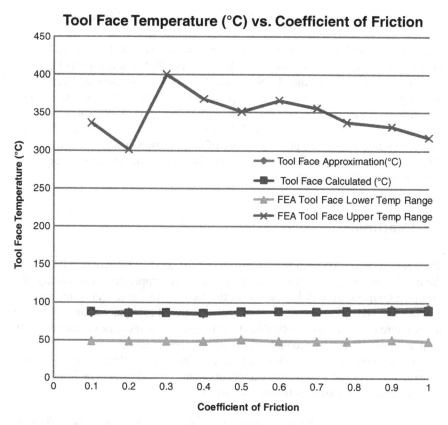

Fig. 2.10 Comparison of approximate and calculated tool face temperatures with finite element generated temperatures at the end of the machining pass (800 mm machined). Re-produced from 'Machining with Nanomaterials, Edited by M. J. Jackson and J. S. Morrell, with kind permission from Springer Science+Business Media B.V. (Permission Received February 22, 2012)

heat towards the shear plane rather than toward the cutting tool. Therefore, during dry machining operations it is necessary to lower the coefficient of friction between chip and tool by providing a thin film coating to the tool in order to lower the coefficient of friction. It is also observed that the tool face temperature and the shear plane temperature increases rapidly as a large chip is cut. The initial stages of chip formation are usually confined to the point at which the chip does not stick to the tool, i.e. perfect sliding takes place between chip and tool. This appears to generate lower temperatures on the tool face. However, once the chip sticks to the tool a secondary shear zone is established and quickly heats up the tool face. It appears that the discrepancy between Loewen and Shaw's calculations and the finite element calculations occurs when adhesion of the chip to the tool takes place, i.e., after the first ten millimeters, or so, of chip formation depending on the coefficient of friction. The calculated peak tool temperature of the initial stage of chip formation is remarkably accurate to the finite element calculations, but not so when the secondary shear zone is established, i.e., when the chip sticks to the tool face.

Figures 2.9 and 2.10 show the difference in values very well at the end of the machining stroke when 300 mm of chip has been cut. When one inspects Figs. 2.3, 2.5, and 2.7, it is clear that Loewen and Shaw's method provides very good accuracy at the primary stages of chip formation. This statement implies that Astakhov's statements concerned with using the early forms of machining mechanics are both valid and invalid depending on how one describes the machining conditions, i.e., primary stages of chip formation where the secondary shear is non-existent, or steady-state machining conditions when the secondary shear zone is fully established.

2.5 Conclusions

The calculations of the change in coefficient of friction have shown that increasing the coefficient of friction will lower the shear plane temperature. Conversely, the tool face temperature will increase with the increase in coefficient of friction.

Loewen and Shaw's method of calculating shear plane and tool face temperature appears to be applicable at the first stages of intimate contact between chip and tool when the secondary shear zone is non-existent.

Loewen and Shaw's method of calculating shear plane and tool face temperature does not appear to be applicable under steady-state conditions when the secondary shear zone is firmly established.

Acknowledgments The author acknowledges permission to reproduce the chapter from the following publication: 'Machining with Nanomaterials, Edited by M. J. Jackson and J. S. Morrell, Chapter 6, 'Analysis of Contact Between Chip and Tool Using Nanostructured Coated Cuttung Tools' by M. J. Jackson, J. Evans. M. D. Whitfield and J. S. Morrell p.p. 199–228, re-printed with kind permission from Springer Science + Business Media B.V. (Permission Received February 22, 2012).

References

1. Astakhov VP, Shvets SV, Osman MOM (1997) Chip structure classification based on mechanisms and their formation. J Mater Process Technol 71:247–257
2. Astakhov VP (1999) Metal cutting mechanics. CRC Press, Florida
3. Astakhov VP (2004) The assessment of cutting tool wear. Int J Mach Tools Manuf 44:637–647
4. Astakhov VP (2005) On the inadequacy of the single shear plane model of chip formation. Int J Mech Sci 47:1649–1672
5. Astakhov VP (2006) Tribology of metal cutting. Elsevier, London
6. Boothroyd G (1961) Photographic technique for determination of metal cutting temperature. Br J Appl Phys 12(5):238–242
7. Bowden FP, Tabor D (1954) The friction and lubrication of solids. Clarendon Press, Oxford

8. Isakov E (2004) Engineering formulas for metalcutting. Industrial Press, New York
9. ISO 8688-2 (1989) Tool life testing in milling—part 2: end milling. International Organization for Standardization, Geneva
10. Loewen EG, Shaw MC (1954) On the analysis of cutting tool temperatures. Trans Am Soc Mech Eng 76:217–231
11. Robinson G (2007) Wear of nanostructured coated cutting tools during mixed scale machining. Ph.D. Dissertation, Purdue University
12. Shaw MC (2005) Metal cutting principles. Oxford University Press, New York
13. Stephenson DA (1991) Assessment of steady-state metal cutting temperature models based on simultaneous infrared and thermocouple data. J Eng Ind 113(2):121–128
14. Wright PK, Trent EM (1973) Metallographic methods of determining temperature gradients in cutting tools. J Iron Steel Inst 211(5):364–368

Chapter 3
Characterization and Use
of Nanostructured Tools

3.1 Introduction

Chemical vapor deposited diamond films have many industrial applications but are assuming increasing importance in the area of microfabrication, most notably in the development of diamond-coated micro-tools especially for milling and turning. For these applications the control of structure and morphology is of critical importance. The crystallite size, orientation, surface roughness, and the degree of sp^3 character have a profound effect on the machining properties of the films deposited. In this chapter experimental results are presented on the effects of nitrogen doping on the surface morphology, crystallite size, and wear of micro-tools. The sp^3 character optimises at 200 ppm of nitrogen and above this value the surface becomes much smoother and crystal sizes decrease considerably.

Fracture induced wear of the diamond grain is the most important mechanism of material removal from a micro-grinding tool during the grinding process. Fracture occurs as a consequence of tensile stresses induced into diamond grains by grinding forces to which they are subjected. The relationship between the wear of diamond coated grinding tools, component grinding forces, and induced stresses in the model diamond grains is described in detail. A significant correlation is found between the maximum value of tensile stress induced in the diamond grain and the appropriate wheel-wear parameter (grinding ratio) machining a selection of engineering steels and cast iron materials. It is concluded that the magnitude of tensile stresses induced in the diamond grain by grinding forces at the rake face is the best indicator of tool wear during the grinding process. Diamond has a unique combination of excellent physical and chemical properties, which makes it ideal for numerous applications [1–3]. It can be used in biomedical components, cutting tools, optical components, microelectronic circuits and thermal management systems. A number of methods have been investigated in order to deposit diamond in thin form to various substrates, the most common being silicon and tungsten carbide cemented with a small amount of cobalt metal [4–6]. Arguably, the most

M. J. Jackson, *Micromachining with Nanostructured Cutting Tools*,
SpringerBriefs in Manufacturing and Surface Engineering,
DOI: 10.1007/978-1-4471-4597-4_3, © The Author(s) 2013

successful method of depositing polycrystalline films of diamond is chemical vapor deposition (CVD). In this chapter we investigate a variant of the basic CVD process known as hot filament CVD for the deposition of diamond. It is generally agreed that the properties of the films, such as morphology, quality, and adhesion that determine the suitability for use in a particular application [7]. In the case of micro-tools, extremely small particles of diamond are required that are blocky in form so that cutting of metals and other materials can be performed with relative ease. Both the diamond nucleation stage and the CVD process conditions critically affect the structure and morphology of diamonds. Abrasion of the substrate material with diamond powder prior to deposition is commonly used to enhance the diamond nucleation density [8–10]. However, such abrasion methods damage the surface in a poorly defined manner. Thus, more controlled methods of nucleation such as biasing prior to CVD are becoming increasingly common [11, 12] and can even enable heteroepitaxial growth of diamond films [13, 14]. The gas-phase environment during deposition also affects the quality and morphology of the resulting diamond films. The addition of nitrogen [15, 16], boron [17] and phosphorus [18] containing gases to the standard methane/hydrogen gas mixtures can change crystal size and its faceting. The effect of changing the stoichiometric balance of the mixture of gases has a significant effect on the development of wear in micro-tools.

Wear mechanisms in micro-tools appear to be similar to that of single-point cutting tools, the only difference being the size and nature of swarf generated. Micro-tools contain very small sharp abrasive grains with blunted cutting edges (known as wear flats), and diamond grains with sharp cutting edges that are released from the grinding tool before they had a chance to remove material from the workpiece. The general form of the wear curve is similar to that of a single-point cutting tool. The wear behavior observed is similar to that found in other wear processes; high initial wear followed by steady-state wear. A third accelerating wear regime usually indicates the occurrence of catastrophic wear of the tool. This type of wear is usually accompanied by thermal damage to the surface of the machined workpiece, which reduces fatigue strength and the life of the component. The performance index used to characterise wear resistance is the grinding ratio, or G-ratio, and is expressed as the ratio of the change in volume of the workpiece ground, ΔV_w, to the change in the volume of the surface of the tool removed, Δv_s, and is shown in Eq. (3.1),

$$G = \Delta v_w / \Delta v_s \qquad (3.1)$$

Grinding ratios cover a wide range of values ranging from less than one for high speed steels [19] to over 60,000 when internally grinding bearing races using cubic boron nitride abrasive wheels [20]. Attempts have been made to address the problems related to the wear of abrasive grains in terms of the theory of brittle fracture [21]. The conclusions of various researchers lead us to believe that the variety of different and interacting wear mechanisms involved, namely, plastic flow of abrasive, crumbling of the abrasive, chemical wear, etc., makes grinding tool wear too complicated to be explained using a single theoretical model [22].

Fig. 3.1 The single-point,
loaded infinite wedge.
Re-produced from
'Machining with
Nanomaterials', Edited by M.
J. Jackson and J. S. Morrell,
with kind permission from
Springer Science+Business
Media B.V. (Permission
Received February 22, 2012)

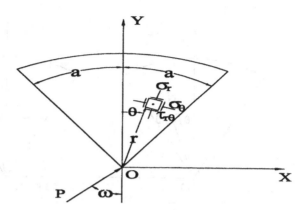

The following analysis of diamond grains represented by single-point loaded wedges assumes that diamond grain fracture is the dominant wear mechanism in a set of grinding tools operating under various grinding conditions. In this chapter it is shown that the addition of nitrogen to methane and hydrogen process gases may be used to influence the surface morphology and structure of the films such that they become suitable for use on micro-tools. The effects of substrate biasing and abrasion on the nucleation and growth of N-doped diamond films, and their influence on the wear of micro-tools is discussed.

3.2 Analysis of Stress in a Loaded Wedge

Diamond grains are angular in habit and possess sharp cutting points prior to grinding workpiece materials. When deposited on a tool substrate, these grains can be considered to be representative infinite wedges. An infinite wedge represents the cutting point of a diamond grain in contact with the workpiece material (Fig. 3.1).

The wedge is loaded at the apex by a load P in an arbitrary direction at angle ω to the axis of symmetry of the wedge. Resolving the force into components P.cos ω in the direction of the axis, and P.sin ω perpendicular to that the stresses due to each of these forces can be evaluated from two-dimensional elastic theory [23]. The state of stress in the wedge, due to force P.cos ω, can be obtained from the stress function,

$$\phi = C.r.\theta.\sin\theta \tag{3.2}$$

where r and θ are polar coordinates at the point N in Fig. 3.2, and C is a constant. The stress function yields the following radial, tangential, and shear stress components,

$$\sigma_r = -2C\frac{\cos\theta}{r} \tag{3.3}$$

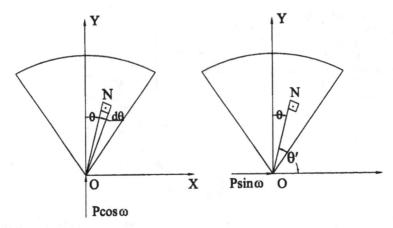

Fig. 3.2 The single-point, loaded infinite wedge showing force components, and the point N within the wedge at polar co-ordinates, r and θ. Re-produced from 'Machining with Nanomaterials', Edited by M. J. Jackson and J. S. Morrell, with kind permission from Springer Science+Business Media B.V. (Permission Received February 22, 2012)

$$\sigma_\theta = 0 \tag{3.4}$$

$$\tau_{r\theta} = 0 \tag{3.5}$$

To determine the constant, C, the equilibrium of forces along the Y-axis is,

$$P \cos \omega - \int_{-a}^{a} \sigma_{r.} \cos \theta dA = 0 \tag{3.6}$$

where dA is an element of cross sectional area within the wedge. If, t, is the thickness of wedge, then,

$$\cos \omega P = \int_{-a}^{a} 2C \frac{\cos \theta}{r} .t.r. \cos \theta d\theta = 2Ct \int_{-a}^{a} \cos^2 \theta d\theta = \text{Ct } (2a + \sin 2a) \tag{3.7}$$

$$\text{Therefore,} \qquad C = \frac{P \cos \omega}{t(2a + \sin 2a)} \tag{3.8}$$

$$\text{And} \qquad \sigma_r = -\frac{2P \cos \theta . \cos \omega}{r.t(2a + \sin 2a)} \tag{3.9}$$

Note that the negative sign denotes that the stress is compressive in this region. The state of stress in the point-loaded wedge, due to force P.sin ω, can be obtained from the following stress function,

$$\phi = C'.r.\theta'. \sin \theta' \tag{3.10}$$

Therefore,

$$\sigma_r = -2C\frac{\cos\theta'}{r} \tag{3.11}$$

$$\text{And,} \quad \sigma_\theta = 0 \tag{3.12}$$

$$\tau_{r\theta} = 0 \tag{3.13}$$

Equilibrium of forces along the X-axis (Fig. 3.2) yields the following solution for the constant, C,

$$P\sin\omega - \int_{\pi/2-a}^{\pi/2+a} \sigma_r.t.r.\cos\theta' d\theta' = 0 \tag{3.14}$$

$$P\sin\omega = -\int_{\pi/2-a}^{\pi/2+a} 2C\frac{\cos\theta'}{r}.t.r.\cos\theta' d\theta' = 2Ct.\int_{\pi/2-a}^{\pi/2+a} \cos^2\theta' d\theta' \tag{3.15}$$

$$= -C.t\,(2a - \sin 2a)$$

$$C = \frac{P\sin\omega}{t(2a - \sin 2a)} \tag{3.16}$$

Thus,

$$\sigma_r = -\frac{2P\cos\theta'.\sin\omega}{r.t(2a - \sin 2a)} \tag{3.17}$$

Expressing in terms of the angle θ (where θ' is negative), yields,

$$\sigma_r = -\frac{2P\cos\theta.\sin\omega}{r.t(2a - \sin 2a)} \tag{3.18}$$

Therefore, the combined stresses are,

$$\sigma_r = -\frac{2P}{r.t}\left[\frac{\cos\omega\cos\theta}{2a + \sin 2a} + \frac{\sin\omega\cos\theta}{2a - \sin 2a}\right] \tag{3.19}$$

It follows that the radial stress, σ_r, vanishes for angle θ_o defined using the expression,

$$\tan\theta_o = \frac{1}{\tan\omega}.\frac{2a - \sin 2a}{2a + \sin 2a} \tag{3.20}$$

This equation corresponds to a straight line through the apex as shown in Fig. 3.3. This natural axis separates the regions of compressive and tensile stresses in the wedge. It can be seen that for values of angle ω which gives, $|\theta_o| > |a|$,

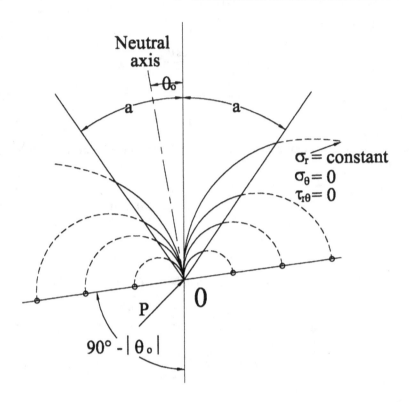

Fig. 3.3 Stress analysis of a single-point loaded wedge. Re-produced from 'Machining with Nanomaterials', Edited by M. J. Jackson and J. S. Morrell, with kind permission from Springer Science+Business Media B.V. (Permission Received February 22, 2012)

provides a neutral axis that lies outside the included angle of the wedge. This means that the whole area of the wedge will be under stresses of uniform sign. Expressing Eq. (3.19) in terms of the rake angle of the diamond grain, β, and force components F_t and nF_t (Fig. 3.3), yields,

$$\sigma_r = -\frac{2F_t}{r.t}\left\{\frac{[n.\cos(a-\beta)+\sin(a-\beta)]\cos\theta}{2a+\sin 2a} + \frac{[\{\cos(a-\beta)-n.\sin(a-\beta)\}\cos\theta]}{2a-\sin 2a}\right\}$$

(3.21)

It is observed that,

$$\tan\omega = \frac{\cos(a-\beta)-n.\sin(a-\beta)}{n.\cos(a-\beta)+\sin(a-\beta)}$$

(3.22)

In the simple case of a wedge with the normal force nF_t along the wedge axis, a is equal to β, hence,

$$\tan\omega = 1/n$$

(3.23)

It is interesting to examine the radial stresses on the left-hand face of the wedge, which corresponds to the leading face of idealised wedge. Thus, for the left-hand face, θ is equal to $-a$, and from Eq. (3.19),

$$\sigma_r = -\frac{2P}{r.t}\left[\frac{\cos\omega\cos a}{2a+\sin 2a} + \frac{-\cos a.\sin\omega}{2a-\sin 2a}\right] \tag{3.24}$$

This stress is zero, i.e., the neutral axis coincides with the left-hand limit of the wedge, when,

$$\frac{1}{\tan\omega} = \frac{\sin a(2a+\sin 2a)}{\cos a(2a-\sin 2a)} \tag{3.25}$$

Thus if,

(a) $a = \beta$, then,

$$n = \frac{\sin a}{\cos a}\cdot\frac{2a+\sin 2a}{2a-\sin 2a} \tag{3.26}$$

(b) $a - \beta = \frac{\pi}{2} - a$ (as is the case when F_t is parallel to the right-hand face of the wedge).

From Eq. (3.23),

$$\tan\omega = \frac{\sin a - n.\cos a}{n.\sin a + \cos a} = \frac{\sin a - n\cos a}{n.\sin a + \cos a} \tag{3.27}$$

And substituting in (3.25), yields,

$$\frac{n\sin a + \cos a}{\sin a - n\cos a} = \frac{\sin a}{\cos a}\cdot\frac{2a+\sin 2a}{2a-\sin 2a} \tag{3.28}$$

Therefore,

$$\frac{\frac{1}{2}.n\sin 2a + \cos^2 a}{\sin^2 a - \frac{1}{2}.n.\sin 2a} = \frac{2a+\sin 2a}{2a-\sin 2a} \tag{3.29}$$

$$\frac{1}{2}.n.\sin 2a(2a-\sin 2a) + 2a\cos^2 a - \cos^2 a.\sin 2a$$

$$= 2a.\sin^2 a + \sin^2 a.\sin 2a - \frac{1}{2}.n.\sin 2a(2a+\sin 2a)$$

$$\frac{1}{2}.n.\sin 2a(2a-\sin 2a + 2a\sin 2a) = 2a(\sin^2 a - \cos^2 a) + \sin 2a$$

$$n.2a.\sin 2a = -2a.\cos 2a + \sin 2a$$

Hence,

$$n = \frac{1}{2a} - \cot 2a \tag{3.30}$$

Equation (3.25) expresses the condition for the whole of the wedge's cross sectional area to be under the influence of a compressive stress. It can be seen that this depends not only upon the rake angle, β, but also upon the force ratio, n. In general the relative size of the region of compressive stresses to the region of tensile stresses depends upon β and n, as Eqs. (3.20) and (3.22) indicate. Also, from Eq. (3.21), the magnitude of the stress on the left-hand face of the wedge is found to be dependent upon the tangential force component, F_t, and the force component ratio, n. Referring to Eq. (3.19), it can be seen that for constant stress, $\sigma_r = $ constant,

$$r.\ C_1 = C_2.\cos\theta + C_3 \cdot \sin\theta \tag{3.31}$$

where C_1, C_2, C_3 are constants. Equation (3.31) represents, in polar co-ordinates, the circumference of a circle tangent to the line. Therefore,

$$C_2.\cos\theta + C_3 \cdot \sin\theta = 0 \tag{3.32}$$

i.e., to the neutral axis at the point when r = 0. However, the point r = 0 must be considered separately because the stress at that point approaches infinity, since by definition P is a point load. The central point of these circles are of constant radial stress, and so the point of constant maximum shear stress must lie on a line perpendicular to the neutral axis at the point where r is equal to zero. The radius of each of those circles depends upon the magnitude of the radial stress, σ_r. Maximum values of stress were computed for a variety of point loads and then correlated to the relevant wear parameter, grinding ratio, for a variety of micro-tools. However, this may not represent the real situation where loads may be distributed along the rake face. Further analysis is required that consider point loads distributed along the rake face.

3.3 Stress Analysis in a Wedge with a Distributed Load

Consider an infinite wedge of included angle, 2a, loaded on one face with a linearly distributed normal and shear load as demonstrated in Fig. 3.4. Within the loaded region of the wedge, the two-dimensional stress components in the wedge can be found using the plane stress function quoted by Timoshenko and Goodier [23] for wedges under polynomial distributed load. In polar co-ordinates, the stress function is,

$$
\begin{aligned}
\phi =& a_0 \log r + b.r^2. \log r + c_0 r^2. \log r + d_0.r^2\theta + d_0'\theta + \frac{a_1}{2}r\theta \sin\theta \\
&+ \left(b_1.r^3 + a_1'r^{-1} + b_1'r \log r\right)\cos\theta - \frac{c_1}{2}r\theta \cos\theta \\
&+ \left(d_1 r^3 + c_1'r^{-1} + d_1'r \log r\right)\sin\theta + \sum_{n=2}^{\infty} \left(a_n r^n + b_n.r^{n+2} + a'.r^{-n} + b_n'.r^{-n+2}\cos n\theta\right) \\
&+ \sum_{n=2}^{\infty} \left(c_n r^n + d_n r^{n+2} + c_{n'}.r^{-n} + d_n'r^{-n+2}\right)\sin n\theta
\end{aligned}
$$

$$\tag{3.33}$$

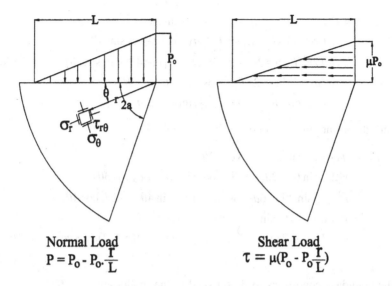

Fig. 3.4 Single-point infinite wedge with a linearly distributed normal and shear load. Re-produced from 'Machining with Nanomaterials', Edited by M. J. Jackson and J. S. Morrell, with kind permission from Springer Science+Business Media B.V. (Permission Received February 22, 2012)

The stress components in the radial, circumferential, and shear directions are,

$$\sigma_r = \frac{1}{r}\frac{\partial \phi}{\partial r} + \frac{1}{r^2}\frac{\partial^2 \phi}{\partial \theta^2}, \qquad \sigma_\theta = \frac{\partial^2 \phi}{\partial r^2},$$

$$\tau_{r\theta} = \frac{1}{r^2}\frac{\partial \phi}{\partial \theta} - \frac{1}{r}\frac{\partial^2 \phi}{\partial r \partial \theta}$$

Considering the terms containing, r^n, and assuming that $n \geq 0$, the radial stress in the wedge is,

$$\begin{aligned}
\sigma_r =\ & 2b_0 + 2d_0 - 2a_2 \cos 2\theta - 2c_2 \sin 2\theta \\
& + r(2b_1 \cos\theta + 2d_1 \sin\theta - 6a_3 \cos 3\theta - 6c_3 \sin 3\theta) \\
& - 12r^2(a_4 \cos 4\theta + c_4 \sin 4\theta) - r^n\{(n^2 - n - 2)(b_n \cos n\theta + d_n \sin n\theta) \\
& + (n+1)(n+2)[a_{n+2}\cos(n+2)\theta + c_{n2}\sin(n+2)\theta]\}
\end{aligned}$$

$$(3.34)$$

Similarly, the circumferential stress in the wedge is,

$$\begin{aligned}\sigma_\theta =& 2b_0 + 2d_0\theta + 2a_2\cos 2\theta + 2c_2\sin 2\theta \\ & + 6r(b_1\cos\theta + d_1\sin\theta + a_3\cos 3\theta + c_3\sin 3\theta) \\ & + 12r^2(b_2\cos 2\theta + d_2\sin 2\theta + a_4\cos 4\theta + c_4\sin 4\theta) \\ & + (n+1)(n+2)r^n[b_n\cos n\theta + d_n\sin n\theta] \\ & + a_{n+2}\cos(n+2)\theta + c_{n+2}\sin(n+2)\theta\end{aligned}$$

(3.35)

And, the shear stress component in the wedge is,

$$\begin{aligned}\tau_{r\theta} =& -d_0 + 2a_2\sin 2\theta - 2c_2\cos 2\theta \\ & + r(2b_1\sin\theta - 2d_1\cos\theta + 6a_3\sin 3\theta - 6c_3\cos 3\theta) \\ & + r^2(6b_2\sin 2\theta - 6d_2\cos 2\theta + 12a_4\sin 4\theta - 12d_4\cos 4\theta) \\ & + r^n[n(n+1)b_n\sin n\theta - n(n+1)d_n\cos n\theta \\ & + (n+1)(n+2)a_{n+2}\sin(n+2)\theta - (n+1)(n+2)d_{n+2}\cos(n+2)\theta]\end{aligned}$$

(3.36)

The boundary conditions to be satisfied in the model are,

$$\sigma_\theta]_{\theta=0} = -\left(P_0 - P_0\frac{r}{L}\right), \sigma_\theta]_{\theta=2a} = 0$$

(3.37)

$$\tau_{r\theta}]_{\theta=0} = -\mu\left(P_0 - P_0\cdot\frac{r}{L}\right), \tau_{r\theta}]_{\theta=2a} = 0$$

(3.38)

Applying the boundary conditions to Eqs. (3.35) and (3.36) yields the following two sets of simultaneous equations,

$$\left.\begin{aligned}& 2b_0 + 2a_2 = -P_0 \\ & 2b_0 + 2d_0.2a + 2a_2\cos 2(2a) + 2c_2\sin 2(2a) = 0 \\ & -d_0 - 2c_2 = -\mu P_0 \\ & -d_0 + 2a_2\sin 2(2a) - 2c_2\cos 2(2a) = 0\end{aligned}\right\}$$

And,

$$\left.\begin{aligned}& 6(b_1 + a_3) = \frac{P_0}{L} \\ & b_1\cos(2a) + d_1.\sin(2a) + a_3\cos 3(2a) + c_3\sin 3(2a) = 0 \\ & -2d_1 - 6c_3 = \mu\frac{P_0}{L} \\ & 2b_1\sin(2a) - 2d_1\cos(2a) + 6a_3\sin 3(2a) - 6c_3\cos 3(2a) = 0\end{aligned}\right\}$$

The simultaneous equations yield the following values for the unknowns,

$$a_2 = \frac{P_0}{4}\left[\frac{(\mu(2a) - 1)\tan^2(2a) + \mu(\tan(2a) - 2a)}{(\tan(2a) - 2a)\tan(2a)}\right]$$

$$b_0 = -\frac{P_0}{2}\left\{1 + \frac{1}{2}\left[\frac{(\mu(2a) - 1)\tan^2(2a) + \mu(\tan(2a) - 2a)}{(\tan(2a) - 2a)\tan(2a)}\right]\right\}$$

$$c_2 = \frac{P_0}{4}\left[\frac{1 - 2\mu(2a) + \mu\tan(2a)}{\tan(2a) - 2a}\right]$$

$$d_0 = \frac{P_0}{2}\left[\frac{\mu\tan(2a) - 1}{\tan(2a) - 2a}\right]$$

$$a_3 = \frac{P_0}{24L}\left[\frac{\tan^2(2a) + 3 - 6\mu\tan(2a)}{\tan^2(2a)}\right]$$

$$c_3 = -\frac{P_0}{24L}\left[\frac{3\mu\tan(2a)(\tan^2(2a) - 1) + 2}{\tan^3(2a)}\right]$$

$$b_1 = \frac{P_0}{8L}\left[\frac{\tan^2(2a) - 1 + 2\mu\tan(2a)}{\tan^2(2a)}\right]$$

$$d_1 = \frac{P_0}{8L}\left[\frac{2 - \mu(\tan^2(2a) + 3)\tan(2a)}{\tan^3 2a}\right]$$

The stresses can now be computed anywhere in the loaded region of the wedge. For the point, r = 0, the angle, θ, is equal to zero, i.e., at the apex of the wedge an arbitrary choice of shear loading on the two faces of the wedge will not produce equilibrium conditions. This is observed if the edge is taken as a 90° corner, and shear loading is considered on one face only. The clearance face is free from shear stress and the corner element is not in equilibrium. Assuming that the error is small, the radial stress, σ_r, at the apex [from Eq. (3.34) and for r = 0, then $\theta = 0$], is,

$$\sigma_r]_{\substack{r=0 \\ \theta=0}} = 2(b_0 - a_2) = -P_0\left[\frac{2a\tan(2a)(\mu\tan(2a) - 1) + \mu(\tan(2a) - 2a)}{(\tan(2a) - 2a)\tan(2a)}\right] \tag{3.39}$$

The analysis shown can be applied to any type of polynomial distribution of the load over the rake face of the wedge-shaped abrasive grain. If the force distribution is not polynomial, an approximate solution can be obtained by considering a linearly distributed load that is equivalent to the original one. Maximum values of stress were computed for a variety of distributed loads and then correlated to the relevant wheel wear parameter, grinding ratio, for a variety of micro-tools.

3.3.1 Development of Wear Model

Brittle materials exhibit high strength properties when loaded in compression than in tension. The ratio of rupture strengths is usually between 3:1 and 10:1 [24]. The existence of relatively low tensile stresses in the diamond grains, may cause failure by fracture to occur. To alleviate fracture, it is possible to use larger grains that reduce the stress levels in the grain when subjected to grinding forces. The effects of changing the amount of nitrogen in the gaseous mixture when creating diamond grains has the effect of changing morphology and the sizes of grains. Therefore, special attention must be paid to deposition conditions that will optimise the size and shape of diamond grains that will resist negative tensile stresses established in the grains when grinding takes place. Therefore, processing conditions determine the life of diamond grains when subjected to large grinding forces. To measure the effectiveness of the deposition process, it is required to model the effects of tensile stresses on the fracture properties of diamond grains deposited in various gaseous environments in order to quantify the effects of nitrogen on the life of micro-tools.

To model the action of the micro-tool, we must consider a single active cutting point to be classed as a wedge of constant width loaded at its inverted apex with point loads, F, and, nF, which represent the radial and tangential force components with reference to the micro-tool in which the grain is supported, and P is the resultant force (Fig. 3.5). The stress distributions within point-loaded wedges can be determined analytically, and the results of such an analysis indicate that if tensile stresses exist within the wedge then it will occur at its maximum along the rake face. The existence of a tensile stress depends on the magnitude of the force ratio, n. If the ratio is especially small that a tensile stress exists in the wedge, then for a specific force ratio the tensile stress is proportional to the tangential grinding force, F. Stresses of this nature would extend to and beyond the diamond grain-substrate interface. The fracture of diamond grain and the interface between the substrate depends on the particular micro-tool used and the magnitude of the tensile stress induced during grinding. Grains of diamond are typically seven-to-ten times stronger in compression than in tension, and therefore the probability of grain fracture is likely to increase with an increase in tensile stress exerted in the grain although the magnitude of the stress may be slightly higher than one-fifth the magnitude of the maximum compressive stress in the grain. A significant barrier to the acceptance of stress patterns evaluated for such situations arises because point loads applied to perfectly sharp wedges produce infinitely high stresses at, and about, the point of contact. Therefore, the loads must be applied over a finite area. This implies that compressive stresses dominate over the finite area. Experimental results [24] show that rake face stresses are compressive over the entire length of chip-tool contact but are tensile outside of this region. The zone of fracture initiation points were located in the tensile zone at about two to two-and-a-half times the chip-tool contact length.

It seems likely that higher tensile stresses are associated with higher grain fracture probability resulting in rapid loss of diamond grains and, consequently, lower grinding ratios. The wear model should incorporate the fact that the loads are spread over a finite area. This implies that single-point loads are resolved into multiple point loads along the rake face, or are indeed, applied directly along the rake face of the cutting tool. The model should allow the relationship between the wear of a micro-tool and the general nature of stresses established in active

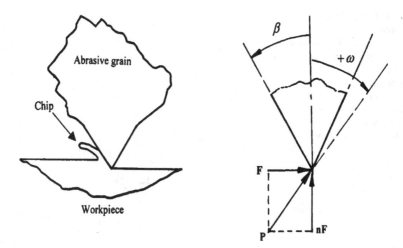

Fig. 3.5 Ideal wedge-shaped cutting point and grinding force diagram. Re-produced from 'Machining with Nanomaterials', Edited by M. J. Jackson and J. S. Morrell, with kind permission from Springer Science+Business Media B.V. (Permission Received February 22, 2012)

diamond grains subjected to grinding forces to be examined. However, the development of a wear model that represents the effects of induced stress in the material is required. The analytical models developed here describe well the initial stages of grinding. However, as the grains are worn by a mechanism of abrasive wear rather than fracture wear, then a computational model must be considered.

3.3.2 Computational Stress Analysis of Single Diamond Grains

The assumed geometry of an ideal grain in the vicinity of its cutting edge is a simple symmetrical wedge of constant width with an included angle of $70°$ that results in a rake angle of $-35°$.

There is no wear flat on the model cutting grain. In order that a finite element method is used to evaluate stresses in the wedge, the wedge was subdivided into 210 diamond-shaped elements with a total of 251 nodes. Forty-one nodes were constrained at the boundary of the wedge and the leading five nodes on the rake face were loaded (Fig. 3.6). The tangential and normal grinding point loads were replaced by a series of multiple loads (F_Y and F_Z) acting perpendicular to (normal load) and along (shear load) the rake face of the wedge.

The loads at the five nodes are representative of the distributed and normal loads acting on the rake face over the diamond grain-chip contact length. The normal force distribution on the rake face was taken as being a maximum value at the cutting edge and decreases linearly to zero at the end of the diamond grain-chip contact length. The shear force was taken to be constant over the first half of the

Fig. 3.6 Finite element
assemblage with grinding
loads applied at the rake face
nodes. Re-produced from
'Machining with
Nanomaterials', Edited by M.
J. Jackson and J. S. Morrell,
with kind permission from
Springer Science+Business
Media B.V. (Permission
Received February 22, 2012)

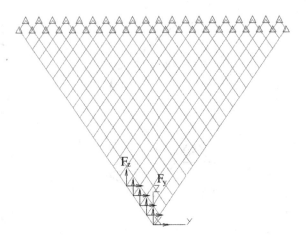

contact length, decreasing linearly to zero over the contact length. Grinding loads
were also applied directly to the rake face and at the tip of the diamond without
applying multiple loads along the rake face. This was performed in order to
compare and contrast the effect of different force distributions on the stresses
generated within the diamond wedge.

To measure the value of using the maximum tensile stress as a way to estimate
grain fracture tendency, the correlation between the two sets of data were calculated for each set of data. The region of fracture initiation was also located using
Griffith's criterion of fracture [25], which is applicable to the fracture of brittle
materials. For,

$$\frac{\sigma_c}{\sigma_t}.\sigma_1 + \sigma_3 > 0 \tag{3.40}$$

Then,

$$\sigma_1 = \sigma_t \tag{3.41}$$

But for,

$$\frac{\sigma_c}{\sigma_t}.\sigma_1 + \sigma_3 < 0 \tag{3.42}$$

Then,

$$(|\sigma_1| - |\sigma_3|)^2 + 8\sigma_t(|\sigma_1| - |\sigma_3|) = 0 \tag{3.43}$$

where σ_1 and σ_3 are the principal stresses, assuming that $\sigma_1 > \sigma_3$, σ_t is the ultimate
tensile strength of the abrasive grain, and σ_c is the ultimate compressive strength.
For diamond, the ratio of σ_t and σ_c is approximately 0.1.

3.4 Experimental Methods

3.4.1 Hot Filament Chemical Vapor Deposition

The hot filament CVD system is composed of a water-cooled stainless-steel vessel, which is connected to a rotary pump enabling a vacuum to be produced. The hot filament CVD apparatus is shown in Fig. 3.7. Gas flow rates are controlled using MKS mass flow controllers to accurately control the amounts of gases flowing into the reactor. The system allows independent bias to be applied between the substrate and filament. The filament consisted of a coiled tantalum wire of diameter 0.5 mm to activate the reaction mixture. The filament temperature was measured using an optical pyrometer with values between 2,200 and 2,500 K. Substrate temperatures were measured using a K-type thermocouple in direct thermal contact with the substrate.

After abrasion, the samples were ultrasonically cleaned with acetone prior to deposition. The diamond films were grown on pre-abraded WC–Co substrates for 4 h under standard deposition conditions [4]. To investigate the effects of changing the gas-phase environment different concentrations of nitrogen, from 50 to 100,000 ppm, were added to the standard 1 % methane in hydrogen gas mixture. This is equivalent to varying the N/C ratio from 0.01 to 20. The film morphology, growth rate, and quality were characterised using Raman spectroscopy (Kaiser Holo Probe: 532 nm, Nd: YAG laser), and scanning electron microscopy (SEM).

3.4.2 Measurement of Wear of Diamond Tools

The measurement of the wear of diamonds on a single layer deposited to the grinding tool requires grinding various workpiece materials on a specially constructed machine tool. The machine tool was constructed using a tetrahedral space frame design that attenuates vibrations generated during grinding. The grinding tool is held in an air turbine spindle capable of rotating the grinding tool in excess of speeds of 350,000 revolutions per minute. Figure 3.8 shows the machine tool complete with three axes of motion in the x, y, and z-directions. A fourth axis capable of rotary motion can also be used on the machine tool. The machine tool was used to measure the performance of the grinding tools machining materials such as medium carbon steels (hypoeutectoid), high carbon tool steels (hypereutectoid), and cast iron. Grinding experiments were conducted using a number of micro-tools coated with diamonds that were produced in a hot filament CVD reactor containing gases with varying amounts of nitrogen in a methane/hydrogen mixture. The grinding ratio was measured in accordance with that stated in Sect. 3.1. However, in order to correlate the magnitude of tensile stress in the diamond grains to the grinding ratio, it is required to know the number of active cutting grains on the surface of the micro-tool.

Fig. 3.7 Hot filament CVD apparatus. Re-produced from 'Machining with Nanomaterials', Edited by M. J. Jackson and J. S. Morrell, with kind permission from Springer Science+Business Media B.V. (Permission Received February 22, 2012)

Fig. 3.8 Micro-machine tool showing tetrahedral spaceframe surrounding the precision x–y–z table and the extremely high-speed air-turbine spindle. Re-produced from 'Machining with Nanomaterials', Edited by M. J. Jackson and J. S. Morrell, with kind permission from Springer Science+Business Media B.V. (Permission Received February 22, 2012)

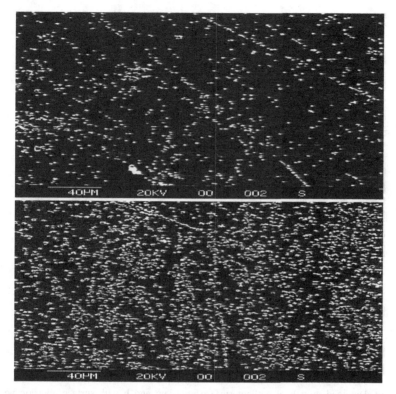

Fig. 3.9 Diamond nucleation on the unabraded surface prior to deposition for 6 h at 0 and 200 ppm nitrogen addition. Re-produced from 'Machining with Nanomaterials', Edited by M. J. Jackson and J. S. Morrell, with kind permission from Springer Science+Business Media B.V. (Permission Received February 22, 2012)

The number of active cutting grains on the micro-tool surface is found quite simply by driving a grinding tool into a piece of soft metal that has a length equal to the grinding tool's circumferential length. The depth to which the grinding tool is driven into the length of soft metal is equal to the depth of cut. The impression that the grinding tool produces in the length of soft metal is equal to the number of cutting points that are active during the grinding stroke at that particular rate of cut. However, the micro-tool must be prepared by simulating the grinding conditions used during the experimental conditions. Once the micro-tool has stabilized at its optimum grinding condition, then the tool is removed from the machine tool and driven into the soft metal that simulates one grinding revolution into the workpiece material. The force components are then applied to a 'model' diamond grain by dividing the grinding force data generated using a dynamometer into the number of active cutting grains over an area that simulates the diamond grain-workpiece contact area over one revolution.

Fig. 3.10 Diamond growth: **a** without surface abrasion; and **b** with surface abrasion with a diamond powder. Re-produced from 'Machining with Nanomaterials', Edited by M. J. Jackson and J. S. Morrell, with kind permission from Springer Science+Business Media B.V. (Permission Received February 22, 2012)

Stresses established in this area are calculated using finite elements. The wear of the micro-tool, expressed in terms of a grinding ratio, and its relationship to the stresses set up in the model grain is compared.

3.5 Discussion

3.5.1 Diamond Deposition

Figure 3.9 shows CVD diamond growth on unabraded WC–Co substrate for 6 h: A striking increase in diamond nucleation density is observed with the presence of nitrogen in the process gas. It is evident that nitrogen enhances the nucleation of diamond. However, even after 6 h it not sufficient on its own to cause the growth of a continuous diamond film. Hence, surface abrasion or another form of substrate preparation is a necessary in the creation of nucleation sites for the growth of diamond films. Figure 3.10 shows CVD diamond growth with and without surface

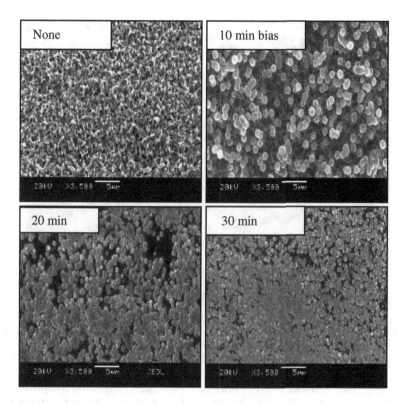

Fig. 3.11 Scanning electron micrograph showing the effects of biasing time on the nucleation density. Re-produced from 'Machining with Nanomaterials', Edited by M. J. Jackson and J. S. Morrell, with kind permission from Springer Science+Business Media B.V. (Permission Received February 22, 2012)

abrasion. It is clearly evident that without abrasion even after 4 h of deposition only isolated crystals of diamond appear on the surface of the substrate. However, surface abrasion results in a continuous film of diamond with the nucleation density of about $9 \times 10^8/cm^2$. Mixed crystals of < 111 > and < 100 > orientation are formed, which are typically 1–3 μm in diameter. The film is continuous with no evidence of pin-holes, or cracks. For micro-tools, a highly controlled method of surface treatment is desirable. Even though surface abrasion is very effective in creating nucleation sites it does not allow a high degree of precision and control of surface preparation and therefore bias enhanced nucleation has been investigated as an alternative.

Application of a negative bias of −300 V to the substrate for 30 min prior to the deposition stage gives a measurable increase in the nucleation density with a continuous film forming (Fig. 3.11). The increase in nucleation is due to the creation of nucleation sites arising from ion bombardment of the substrate. However, at the pressures utilised in this study of 2,660 Pa the mean free path is relatively small and ion acceleration appears unlikely and an alternative

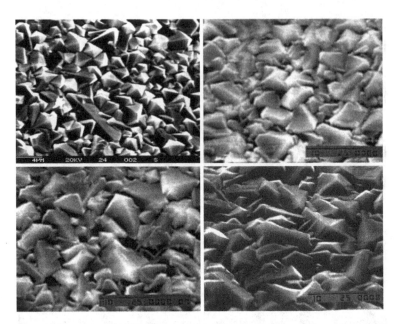

Fig. 3.12 N-doped CVD diamond growth at 0, 50, 100 and 200 ppm of nitrogen. Re-produced from 'Machining with Nanomaterials', Edited by M. J. Jackson and J. S. Morrell, with kind permission from Springer Science+Business Media B.V. (Permission Received February 22, 2012)

explanation may be more likely. Biasing the substrate changes the composition of the plasma creating a higher concentration of hydrogen radicals near to the substrate and therefore the changes in the morphology could be due to a chemical etching effect rather than an ion bombardment effect. It is evident from Fig. 3.12 that as the amount of nitrogen is increased from 0 to 200 ppm the size of diamond crystals also increases. The addition of nitrogen is thus enhancing the growth of the diamond crystals. However, the precise reasons for this growth enhancement are still unclear and are being investigated. Figure 3.13 illustrates that further additions of nitrogen degrade the crystal structure considerably but improves the surface roughness of the films.

These films were grown under the same conditions but in separate studies. The different grain densities indicate the difficulty in obtaining absolute reproducibility, though the relative density measurements are significant. Raman analysis confirms this trend. For relatively small amounts of nitrogen in the gas phase the FWHM of the $1,332$ cm^{-1} absorption characteristic of diamond is reduced indicating an increase in the diamond phase purity. For higher levels of nitrogen the diamond peak broadens and disappears altogether with N_2 levels above 50,000 ppm in the gas mixture (Fig. 3.14). This data is consistent with films produced from microwave diamond CVD studies [8]. The changes in the surface morphology and structure are related to the carbon supersaturation, which is controlled by the supply of carbon and the creation of growth sites at the surface [8]. Small amounts of nitrogen are able to reduce carbon supersaturation, which leads to an improvement

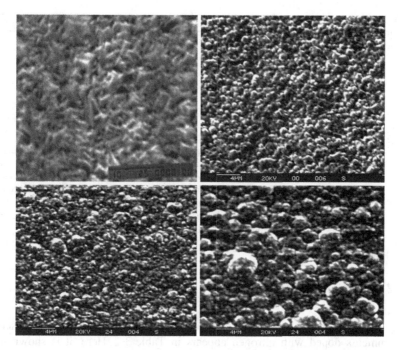

Fig. 3.13 N-doped diamond at 5,000, 10,000, 50,000 and 100,000 ppm of nitrogen. Re-produced from 'Machining with Nanomaterials', Edited by M. J. Jackson and J. S. Morrell, with kind permission from Springer Science+Business Media B.V. (Permission Received February 22, 2012)

Fig. 3.14 Raman spectra of n-doped diamond coatings: **a** 200 ppm; **b** 100 ppm; **c** 5,000 ppm; **d** 50 ppm; **e** 10,000 ppm; **f** 50,000 ppm; and **g** 0 ppm. Re-produced from 'Machining with Nanomaterials', Edited by M. J. Jackson and J. S. Morrell, with kind permission from Springer Science+Business Media B.V. (Permission Received February 22, 2012)

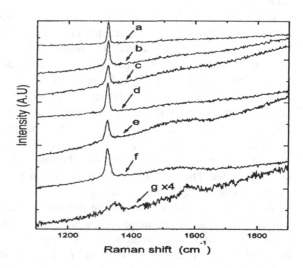

in crystal quality. At higher levels of nitrogen the carbon supersaturation increases and the morphology changes from blocky to spherical, which is accompanied by deterioration of the diamond phase purity.

Table 3.1 Performance data for grinding normalized M2 tool steel

Nitrogen (ppm in methane/hydrogen mixture)	Grinding ratio
0	50
50	110
100	200
200	300
5,000	175
10,000	85
50,000	60
100,000	20

Grinding conditions: *tool diameter* 750 µm, *spindle speed* 350,000 rpm, *depth of cut* 10 µm, *feed rate* 10 µm/min. Re-produced from 'Machining with Nanomaterials', Edited by M. J. Jackson and J. S. Morrell, with kind permission from Springer Science+Business Media B.V. (Permission Received February 22, 2012)

3.5.2 Wear of Diamonds

The wear of micro-tools coated with diamonds deposited from various gaseous environments doped with nitrogen appears in Table 3.1. Here, it is shown that the nature of the diamond is optimised at 200 parts-per-million of nitrogen in methane/hydrogen mixture. Beyond this value, the grinding ratio decreases because the diamonds are becoming smaller in size and the surface is becoming smoother.

Below 200 ppm, the diamonds are scattered in random formation producing a discontinuous film of diamonds leaving large areas of the tool unable to grind the surface of the workpiece material. The results of the two-dimensional stress analyses were consistent with the experimentally determined stress distribution obtained by Loladze [26] when cutting soft metal with photoelastic tools. The maximum tensile stress always occurs at the rake face at a distance from the cutting edge ranging from 1.5 to 4 times the abrasive grain-chip contact length, the exact magnitude of the coefficient depends on the loading conditions for a particular grinding event. For a given value of the tangential force component, F, the higher the force ratio, F/nF, the greater the distance the maximum tensile stress is away from the cutting edge. These results indicate that mechanically induced fracture occurs at a finite distance away from the cutting edge.

When using Griffith's criterion, the influence of mechanically induced stresses indicate that fracture initiation zones are established. Figure 3.15 shows the occurrence of such zones in an idealized wedge. The first zone is located around the point of maximum tensile stress and is always at the rake face.

Failure in this zone is tensile and would initiate fracture at a point on the rake face of the order of two-to-three times the diamond grain-chip contact length away from the cutting edge. This type of fracture is consistent with fracture on a scale

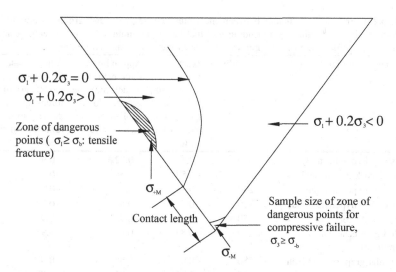

$\sigma_1 + 0.2\sigma_3 = 0$

$\sigma_1 + 0.2\sigma_3 > 0$

Zone of dangerous
points ($\sigma_1 \geq \sigma_b$: tensile
fracture)

$\sigma_1 + 0.2\sigma_3 < 0$

σ_{+M}

Contact length

Sample size of zone of
dangerous points for
compressive failure,
$\sigma_3 \geq \sigma_{-b}$

σ_{-M}

Fig. 3.15 Griffith's criterion applied to the idealized wedge showing tensile and compressive fracture initiation zones. Re-produced from 'Machining with Nanomaterials', Edited by M. J. Jackson and J. S. Morrell, with kind permission from Springer Science+Business Media B.V. (Permission Received February 22, 2012)

comparable with the chip thickness and tends to produce the so-called 'self-sharpening action'. The second much smaller zone is located at the immediate vicinity of the cutting edge. Failure is compressive in this region and results in 'crumbling' of the cutting edge leading to the formation of a wear flat on the diamond grain.

The correlation between the magnitude of the maximum tensile stress in the model diamond grains and the grinding ratio (Table 3.2) is high and is dependent on the way the forces are applied to the grains. It would be expected that the higher the tensile stress, the greater is the rate of grinding wheel wear and consequently the corresponding grinding ratio. Perfect linear correlation in accordance with this would result in a correlation coefficient of -1. The correlation coefficient between the maximum tensile stress and the grinding ratio is significant. This is to be expected as the force ratio may vary slightly. However, if the tangential component of the grinding force changes significantly without a change in force ratio, then it is expected that the maximum tensile stress will change significantly and consequently reduce the grinding ratio.

The calculation and application of multiple grinding loads along the rake face produces a lower correlation coefficient compared to directly applied grinding loads. This implies that grinding loads are simply not point loads acting act the tip of the inverted apex and along the abrasive grain-chip contact length of the grinding grain. In fact, directly applied grinding forces produce better correlation coefficients. This means that for perfectly micro-tools, one must apply the component grinding loads directly to the rake face.

Table 3.2 Correlation coefficient between maximum tensile stress and grinding ratio for an idealized wedge that simulates a grinding tool that has an optimum diamond coverage using a controlled gas atmosphere of 200 ppm of nitrogen in methane/hydrogen mixture

Workpiece material	Grinding ratio	Exact wedge model with point loads applied to apex of the wedge	Approximate finite element model: multiple grinding loads applied to rake face of wedge	Approximate finite element model: grinding forces applied directly to the rake face of the wedge
En2 steel (hypoeutectoid)	500	−0.7	−0.82	−0.99
Normalized M2 tool steel	300	−0.54	−0.68	−0.98
En8 steel (hypoeutectoid)	250	−0.15	−0.25	−0.6
AISI 52100 (hypereutectoid)	350	−0.76	−0.86	−0.98
Annealed M2 tool steel	250	−0.8	−0.9	−0.99
Chilled grey cast iron (flake graphite)	650	−0.83	−0.85	−0.91
Spheroidal graphite cast iron	580	−0.9	−0.95	−0.99
Austempered ductile iron (bainitic structure)	525	−0.92	−0.95	−0.99

Comparison is also made between the methods of applying loads to the idealized wedge models. Re-produced from 'Machining with Nanomaterials', Edited by M. J. Jackson and J. S. Morrell, with kind permission from Springer Science+Business Media B.V. (Permission Received February 22, 2012)

3.6 Conclusions

We have shown that the surface morphology of the diamond films can be controlled by employing surface abrasion, substrate biasing, or nitrogen addition to the gas mixture. The application of a bias voltage during normal diamond growth enables re-nucleation of the diamond film. Thus, the crystal size and surface roughness may be controlled with no reduction in the diamond phase purity. Addition of nitrogen to the gas phase during diamond CVD can also be used to control the surface morphology. For diamond-coated microtools, grain fracture appears to be the dominant cause of diamond loss during a grinding operation. Grain fracture is much more likely to be caused by mechanically induced tensile stresses within diamond grains than by mechanically induced compressive stresses. The best indicator of microtool performance during a grinding operation under different operating conditions is the level of tensile stress established in diamond grains. High tensile stresses are associated with grain fracture and low grinding ratios in perfectly sharp micro-tools. Finite element models of sharp diamond grains can be applied to micro-tools where the dominant wear mechanism is grain fracture. The development in formulating better diamond coated micro tools is of utmost importance to creating efficient machining at the microscale.

Acknowledgments The author acknowledges permission to reproduce the chapter from the following publications: 'Machining with Nanomaterials', Edited by M. J. Jackson and J. S. Morrell, Chapter 10, 'Manufacture and Development of Nanostructured Diamond Tools' by M. J. Jackson, W. Ahmed, and J. S. Morrell pp. 325–360, re-printed with kind permission from Springer Science+Business Media B.V. (Permission Received February 22, 2012), and 'Characterization from N-doped polycrystalline diamond films deposition to micro tools', by M. J. Jackson and W. Ahmed, Journal of Materials Engineering and Performance, Vol. 14 (5), 2005, pp. 654–665 (Springer License# 2944340724659—Issued 8th July 2012).

References

1. Fan QH, Periera E, Gracio J (1998) Time modulated CVD diamond processing of diamond. J Mater Res 13(10):2787–2794
2. May P, Rego CA, Thomas RM, Ashfold MNR, Rosser KN, Everitt NM (1994) CVD diamond wires and tubes. Diam Relat Mater 3:810–813
3. Kostadinov L, Dobrev D, Okano K, Kurosu T, Iida M (1992) Nuclear and growth of diamond from the vapor phase. Diam Relat Mater 1:157–160
4. Ali N, Ahmed W, Hassan IU, Rego CA (1998) Surface engineering of diamond coated tools. Surf Eng 14(4):292
5. Beckmann R, Kulisch W, Frenck HJ, Kassing R (1992) Influence of gas phase parameters on diamond kinematics of thin diamond films deposited by MWCVD and HFCVD techniques. Diam Relat Mater 1:164–167
6. Ojika SI, Yamoshita S, Ishikura T (1998) Diamond growth on copper substrate. Jpn J Appl Phys 32(2):L1681–L1683
7. Muller-Serbert W, Worner E, Fuchs F, Wild C, Koidl P (1996) Nitrogen induced increase in growth rate in CVD diamond. Appl Phys Lett 68(6):759–760
8. Bohr B, Haubner R, Lux B (1996) Influence of nitrogen additions on HFCVD diamond. Appl Phys Lett 68(8):1075–1077
9. Yarbrough WA, Messier R (1988) Diamond deposition to silicon. Science 247:688
10. Kanetkar SM, Metera G, Chen X, Pramanick S, Tiwari P, Narayan J, Pfeiler G, Paesler M (1991) Growth of diamond on silicon substrates. J Elect Mater 20:4
11. Wolter SD, Stoner BR, Glass JT, Ellis PJ, Jenkins DS, Southworth P (1993) Textured growth of diamond on silicon via in situ carburisation and bias enhanced nucleation. Appl Phys Lett 62:1215–1217
12. Jiang X, Klages CP, Zachia R, Hartureg M, Fuser HJ (1993) Epitaxial diamond films on (001) silicon substrates. Appl Phys Lett 62:3438–3440
13. Stubhan F, Ferguson M, Fusser HJ, Behom RJ (1995) Heteroepitaxial nucleation of diamond on Si (001) in HFCVD. Appl Phys Lett 66:1900–1902
14. Li X, Hayashi Y, Nishino S (1997) Analysis of oriented diamond nucleation processes on silicon substrates by HFCVD. Jpn J Phys 36:5197–5201
15. Locher R, Wild C, Herres N, Behr D, Koidl P (1994) Nitrogen stabilized <100> texture in CVD diamond films. Appl Phys Lett 65:34–36
16. Jin S, Moustakas TD (1994) Effect of nitrogen on the growth of diamond films. Appl Phys Lett 65:403–405
17. Borst TH, Munzinger PC, Weiss O (1994) Characterization of undoped and doped homoepitaxial diamond layers produced by microwave plasma CVD. Diam Relat Mater 3:515–519
18. Koizumi S, Kamo M, Sato Y, Mita S, Sauabe A, Reznik C (1998) Growth and characterization of phosphorous doped n-type diamond films. Diam Relat Mater 7:540–544
19. Tarasov LP (1951) Grindability of tool steels. Am Soc Metals 43:1144–1151

20. Tonshoff HK, Grabner T (1984) Cylindrical and profile grinding with boron nitride wheels. Proceedings of the 5th international conference on production engineering, Japanese Society of Precision Engineers, p 326
21. Malkin S, Cook NH (1971) The wear of grinding wheels—Part 1: attritious wear. Trans ASME J Eng Ind 93:1120–1128
22. Jackson MJ (2001) Vitrification heat treatment during the manufacture of corundum grinding wheels. J Manuf Process 3:17–28
23. Timoshenko SP, Goodier JN (1970) Theory of elasticity, 3rd edn.—International Student Edition. McGraw-Hill Kogakusha Ltd., New York, pp 109–113 and 139–144
24. King AG, Wheildon WM (1966) Ceramics in machining processes. Academic Press, New York
25. Griffith AA (1921) The phenomena of rupture and flow in solids. Philos Trans R Soc Lond A221:163–198
26. Loladze TN (1967) Requirements of tool materials. Proceedings of the 8th international machine tool design and research conference. Pergamon Press, pp 821–842

Index

M. J. Jackson, *Micromachining with Nanostructured Cutting Tools*,
SpringerBriefs in Manufacturing and Surface Engineering,
DOI: 10.1007/978-1-4471-4597-4, © The Author(s) 2013